Strategy and Partnership in Cities and Regions

Also by Brian Jacobs

BLACK POLITICS AND URBAN CRISIS IN BRITAIN

FRACTURED CITIES: Capitalism, Community and Empowerment in Britain and America

PUBLIC–PRIVATE PARTNERSHIPS FOR LOCAL ECONOMIC DEVELOPMENT (*editor with N. Walzer*)

RACISM IN BRITAIN

Strategy and Partnership in Cities and Regions

Economic Development and Urban Regeneration in Pittsburgh, Birmingham and Rotterdam

Brian Jacobs
Professor of Public Policy
School of Health
Staffordshire University

First published in Great Britain 2000 by
MACMILLAN PRESS LTD
Houndmills, Basingstoke, Hampshire RG21 6XS and London
Companies and representatives throughout the world

A catalogue record for this book is available from the British Library.

ISBN 978-0-333-77782-4

First published in the United States of America 2000 by
ST. MARTIN'S PRESS, INC.,
Scholarly and Reference Division,
175 Fifth Avenue, New York, N.Y. 10010

ISBN 978-1-349-62791-2 ISBN 978-1-137-05184-4 (eBook)
DOI 10.1007/978-1-137-05184-4

Library of Congress Cataloging-in-Publication Data
Jacobs, Brian D. (Brian David), 1950–
Strategy and partnership in cities and regions : economic development and urban regeneration in Pittsburgh, Birmingham and Rotterdam / Brian Jacobs.
p. cm.
Includes bibliographical references and index.
ISBN 0–312–23028–1 (cloth)
1. Urban renewal—Pennsylvania—Pittsburgh. 2. Urban renewal—England—Birmingham. 3. Urban renewal—Netherlands—Rotterdam. 4. Public–private sector cooperation—Pennsylvania—Pittsburgh. 5 Public–private sector cooperation—England—Birmingham. 6. Public–private sector cooperation—Netherlands—Rotterdam. I. Title.

HT177.P5 J33 1999
307.3'416—dc21

99–048148

Softcover reprint of the hardcover 1st edition 2000 978-0-312-23028-9

This book is printed on paper suitable for recycling and made from fully managed and sustained forest sources.

10 9 8 7 6 5 4 3 2 1
09 08 07 06 05 04 03 02 01 00

To my wife and son, Helen and Mark

Contents

List of Figures

List of Maps

Acknowledgements

I wish to thank the following people who granted interviews and provided information during the course of my research. The views and interpretations in the book are entirely my own and in no way should be taken to be representative of any of the individuals or organisations mentioned below.

For the Birmingham case, interviews were conducted with C. Brockie, S. Dickinson, L. Lillis, and S. Taylor, of Birmingham City Council; S. Dodds and H. Duffy, Birmingham Heartlands Development Corporation; P. Jeffrey, ECOTEC, Birmingham; D. Mahoney, Government Office for the West Midlands; D. Prior, Birmingham City Pride, and T. Setchell, Saltley-Small Heath Partnership. I am grateful for the assistance provided by C. Dutton OBE, advisor to the British government's Urban Task Force and Head of Regeneration at Sandwell Metropolitan Council: I also thank Aston–Newtown City Challenge in Birmingham and the British Urban Regeneration Association, London.

For the Rotterdam case, interviews were conducted with Professor L. Cachet and Professor J. Koppenjan at Erasmus University; J. Dullaart, BOF; T. Geerts and K. W. Vrijdag, Chamber of Commerce and Industry for Rotterdam and the Lower-Maas; S. Jonker and P. van der Reijden, City of Rotterdam; Dr A. Kerkum-Hubregtse and P. Rodenberg, Kop van Zuid Project Team; J. Leemhuis, Global Business Network (GBN), Amsterdam; J. van Boven and K. Machielse, OBR, Rotterdam (for use of material on the OBR–GBN scenario exercise); J. van der Meer, European Institute for Comparative Urban Research; G-J. Meijer, Kop van Zuid Social Return Project, and C. J. Houterman at the Ministry of Internal Affairs. I am also grateful for the assistance provided by Professor T. Toonen, at Leiden University; Professor P. Drewe, Delft University of Technology; Dr F. Hendricks, Tilburg University; Professors A. Kreukels and M. Bovens, Utrecht University; the International Urban Development Association, The Hague, and Amsterdam City Housing Department.

For the Pittsburgh case, interviews were conducted with D. Black and S. Phillips, Pittsburgh Partnership for Neighborhood Development; Dr M. Birru, Urban Redevelopment Authority of Pittsburgh; D. Gray, US Department of Housing and Urban Development, Pittsburgh; Professor R. Stafford, Allegheny Conference for Community Development (ACCD), and P. K. Larson at the Southwestern Pennsylvania Regional Planning Commission. I am grateful for the assistance provided by Professors R. Bangs, M. Coleman, L. Comfort, S. Deitrick, G. Peters, at the University of Pittsburgh; Professor J. Tarr, Carnegie Mellon University; R. Bradwein, Policy and Management Associates, Boston; W. Dodge, Strategic Partnerships Consulting; M. Fanzo, Independent Writer; Professor D. Smith, Carnegie Mellon University for information about regional 'white paper'; the Steel Industry Heritage Corporation; the Mayors Office of the City of Pittsburgh; the Pittsburgh–Allegheny Enterprise Community, and the Ford Foundation.

Thanks go to the following for the use of material: Westview Press for adaptation of presentation in Figure 1.1; US Census Bureau, Figure 4.1; ACCD, Pittsburgh, Figure 4.2; The Office for National Statistics, Figure 6.1; Birmingham City Council, Figure 7.1; City of Rotterdam and CBS, Figure 8.1; OBR Rotterdam, for translated adaptations of material in Figures 8.2, 8.3 and 8.4.

I thank Rosemary Duncan at Staffordshire University who drew the maps. A special thanks goes to Professor Paul 't Hart and Marieke Kleiboer of Leiden University who hosted me during one of my research visits to the Netherlands. They, and their colleagues, guided me to many key sources and contacts in the Netherlands, and elsewhere. Finally, thanks to Alison Howson and Keith Povey for their work on the production of the book and their patient handling of my numerous revisions and queries.

BRIAN JACOBS

Introduction

This book deals with competitive policy strategies in the urban regions of Pittsburgh (USA), Birmingham (United Kingdom) and Rotterdam (the Netherlands). The cases included deal with innovative strategies to deal with economic restructuring and the promotion of the regional interests of cities. They are about the ways partnerships link governments, corporations, and community groups in economic development and urban regeneration. The book provides a model, locating partnerships and big city governments in the context of complex changes in the policy environment. The research concerns the nature of partnerships and the relationship between strategy and partnership structure. Therefore, the book does not deal in depth with matters of political leadership or with the inner workings of local government strategy making processes. It is not about committee or council leadership wrangles or the influence of civic leadership styles. While such issues are of importance they are outside the scope of the present work. The approach here is to deal with the contexts and structures that influence partnership and the organisational aspects of public–private collaboration.

New patterns of world trade, the decline of staple industries, and profound social and demographic changes are reordering economic life. The growing 'knowledge driven economy' (Department of Trade and Industry, 1998) speeds economic change as corporations forge strategic alliances and as urban regions reposition themselves in the market to compete. Urban policy makers deal with the consequences of global and regional change as industries restructure and companies relocate. Corporations trade globally, but they also become embedded in domestic regional economies where they have production plants and headquarters (Amin, 1994). The social disruption caused by corporate downsizing, reengineering and the shift from manufacturing to services creates severe problems as European and North American industries restructure. Transnational corporations join regional alliances and networks, and in the three case study cities, public officials and business leaders think strategically effectively to respond to such changes. Policy makers recognise the importance of the core economic strengths of their regions and they know that competitiveness depends upon skilled citizens living in prosperous communities. They know that the most successful policies address the problems of urban communities and that competitive policies lead cities to work regionally through networks often removed from traditional local government. Public–private partnerships can be flexible and innovative, but they redefine the links between local governments, corporations, and community groups by challenging traditional structures and processes. Nevertheless, in economic development, city governments have assumed important new roles as they participate in policy networks and partnerships.

Why Pittsburgh, Birmingham and Rotterdam?

The European Commission (1998, 1999) has advocated the comparative policy analysis of changing regional economies in the European Union and has compared policies with those in the USA (European Commission, 1995a). European Union policy, with its developing urban agenda, is emphasising the connections between cities across borders through new alliances and partnerships as cities contribute to economic growth as dynamic hubs in regional economies. The British Labour government's Urban Task Force recently studied urban regeneration in North American and European cities, and other studies have demonstrated the value of comparison (Harding, Dawson, Evans and Parkinson, 1994). Comparative approaches require the careful selection of case studies, so criteria were adopted in the selection of the cities in this book that stipulated that the subjects should broadly be comparable as large 'central cities'. They would be in regions with fast restructuring economies with regional policy innovation and active public–private partnerships. They were likely to be cities that had adopted holistic policies to regenerate declined communities. The following explains how the selected cities satisfied these criteria.

Pittsburgh, Birmingham and Rotterdam were selected because they are broadly comparable within the international urban hierarchy as important 'urban centres'. Global competition between cities means that some cities grow rapidly while others falter with big cities battling to retain or improve their positions in the global hierarchy. Friedmann (1995) defines cities according to their functions in the global, national and regional economies. 'World cities' are important at each spatial level, and they usually have between one and twenty million inhabitants. World cities affect the economies of other cities and act as financial and commercial hubs. Friedmann (1995, p. 23) classifies world cities within a 'hierarchy of spatial articulations' where many variables determine the global and regional roles of cities. Some cities have more global economic influence than others do, and they influence their regions to different degrees. Friedmann (1995, p. 23) argues that it may be a 'futile undertaking' to define a 'fixed and stable' hierarchy of world cities because there are no neat categories. Nevertheless, major world cities such as London, Tokyo, and New York are command centres in the hierarchy because they 'articulate' regional, national and other international economies into the global economy. They do this by providing important financial services and by way of companies with leading roles in world markets. Other world cities such as Los Angeles, Amsterdam, Singapore and Frankfurt produce 'multinational articulations' (Friedmann, 1995, p. 23) acting as centres controlling large global capital flows. However, the differences between multinational cities and other world cities such as Paris, Zurich, Sydney, and Madrid that are 'important national articulations' are less clear (Friedmann, 1995, p. 24). Paris and Madrid integrate their respective national economies into the international system, and they are important globally through their financial and commercial services. Friedmann's 'subnational–regional articulations' are not easily distinguished from the 'national articulations', but the subnational cities rely more on their regional contributions to national and international economies. Typically, these are cities such as Lyon, Barcelona, Boston and Munich. Here, it is argued that Pittsburgh, Birmingham and Rotterdam dominate their regions, but they are not world cities. They do not conform to Hall's (1995) subglobal cities, such as Barcelona and Zurich, nor do they fully conform to Friedmann's global nodes. Instead, Pittsburgh, Birmingham and Rotterdam are important cities below the big global, multinational and subnational world cities. It is true that to different degrees, the three case study cities integrate regional economies into national and global economies, but they lack the range of institutions, cultural attractions, large populations, economic power and overall global salience of world cities.

Pittsburgh, at the heart of Southwestern Pennsylvania (see Map I.1), had an estimated population of 350,363 in 1996 with an estimated 2,379,411 in the wider Pittsburgh Metropolitan Statistical Area (MSA). The city serves a large regional industrial sector and it also relies upon services and high technology. Pittsburgh has an important global presence with its Fortune 500 corporate headquarters, but it has to compete with larger American cities. Rotterdam and

Birmingham are second most populous national cities with Rotterdam having a population of 598,694 in 1995 and 1,143,648 in its wider metropolitan area. Birmingham had a population of 1,017,000 with 5,306,000 in the surrounding West Midlands region (see Map I.2) in 1995. Like Pittsburgh, policy makers in Rotterdam and Birmingham boast that their cities are centres of international industry, commerce and culture. Rotterdam (see Map I.3) is part of the wealthy Randstad (Ring City) that makes up an economically vital part of the Dutch national economy. Of the three case study cities, Rotterdam probably comes closest to world city status, being the world's largest seaport and a major centre for petrochemicals, commerce and corporate services. However, Friedmann excludes Rotterdam from his list of world cities because Amsterdam dominates the financial services sector within the Randstad and competes with Rotterdam as a national articulation. Rotterdam and Birmingham both rely upon industrial, commercial and service activities to sustain employment and growth but, unlike Amsterdam or London, they are not economic and financial command centres within their own national economies.

The Randstad, which includes Rotterdam, Amsterdam, the Hague and Utrecht, forms one of the most important and wealthy core regions in the European Union. Despite this, the major cities within that region retain their own distinctive identities. Similarly Birmingham is part of the wider West Midlands that includes Coventry, Wolverhampton and Dudley, but these towns and cities take pride in their independence. In practice, therefore, the term 'regional' is a coverall that takes no account of the real diversity of regions. The north American 'urban region', the British 'region' and the Dutch 'city province' signify variety and they are added to by the different regional perceptions of partnership and public agencies. Barnes and Ledebur (1998) argue that urban regions in north America consist of central cities such as Pittsburgh that economically connect to hinterlands that are often defined as MSAs. For Barnes and Ledebur, the United States consists of regional economies that all differ in size and in their relationships to state governments, so sometimes substate level economic regions do not fit local government jurisdictions. The urban region concentrates production in areas around one or more central cities creating agglomeration economies that attract companies (Scott, 1998) and produce regional industry clusters. Regional definitions based on economic areas therefore cut across local government jurisdictions so that 'functional

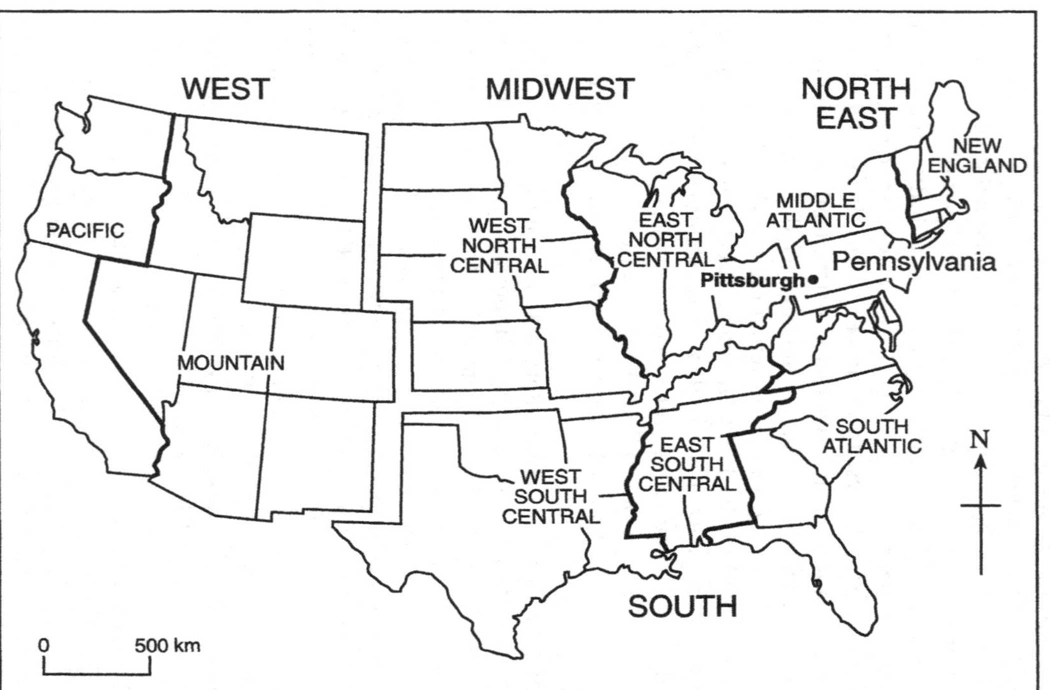

Map I.1 Exploded Map of the United States of America showing Census Regions, Pennsylvania and Pittsburgh

Map I.2 Boundaries of the English Government Offices for the Regions Showing Birmingham

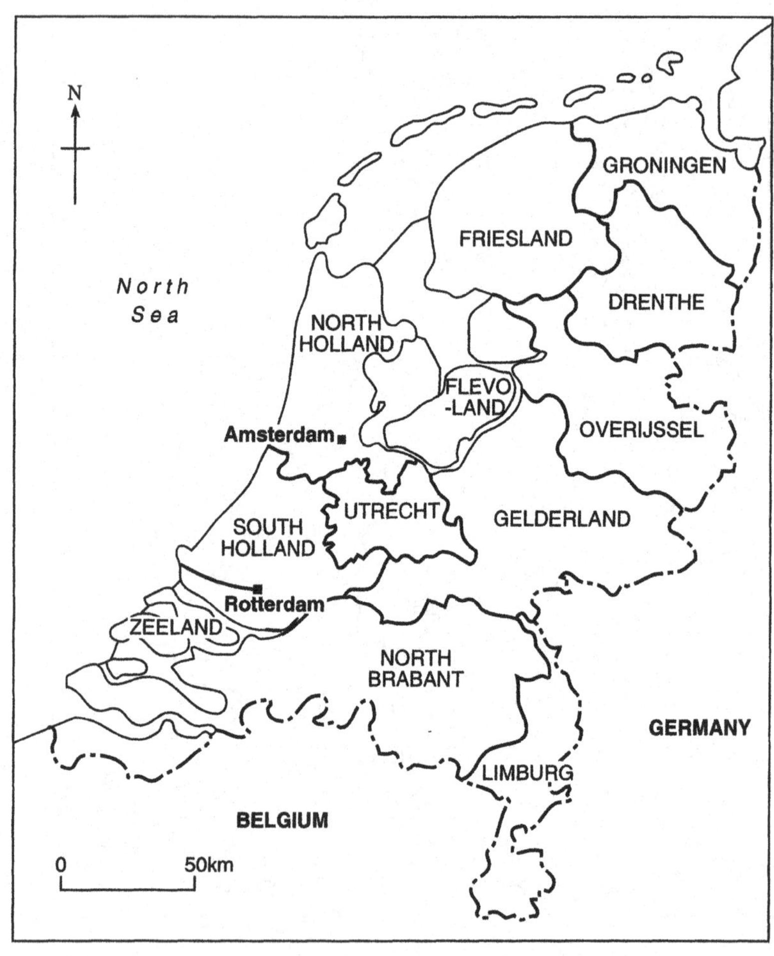

Map I.3 The Netherlands Showing the Provinces and Rotterdam

economic areas do not correspond to any existing local government jurisdictions' (Barnes and Ledebur, 1998, p. 22). Le Gales and Lequesne (1998) refer to similar problems in the European Union where economic regions do not conform to political regions, and varied conceptions of the region reflect changing patterns of economic activity. Different agencies and partnerships have different definitions of regions because those definitions suit their particular operational requirements. This suggests that the 'region' denotes variability in organisational meaning that is symptomatic of the complexity of modern governance. This adds value to the economic region in the Barnes and Ledebur (1998) sense of a central city and surrounding counties, as in Pittsburgh, where there are economic interactions between city and hinterland that produce organisational complexity and variety. Birmingham, a central city in the American sense, is surrounded by counties and large towns for which it provides financial, industrial and commercial services. The 'central city' of Rotterdam, and small surrounding local municipalities, constitute an 'urban region' that is compact and more akin to an American contiguous urban area than to an MSA. However, Rotterdam economically orientates to the Randstad and global markets through its World Port and by the presence of large transnational corporations.

The cases show that global economic restructuring creates important external environmental challenges for policy makers. The external environments of cities are risky and uncertain. Public policy makers have to deal with factory closures, unemployment, and changing social conditions and attitudes. Economic restructuring hastens social change as old industries decline in the post-industrial economy. For example, the Pittsburgh steel industry shrank in the 1980s as politicians fought for the survival of communities. The old identity of Pittsburgh with smokestack industry and militant labour politics gave way to the high technology service economy, single-issue politics and community activism. In Birmingham, militant labour politics declined as old industries faltered in the 1970s and the service sector of the economy grew. In Rotterdam, the city council has responded to the global competitive challenge by reassessing the city's relationship to the world economy and fostering public–private partnerships in urban regeneration and local service delivery. The selected cases therefore present three cities that have had to adjust their economies to remain competitive and to take advantage of the opportunities arising from new technologies and the growth of services. Such hard adjustments contrast with rapidly expanding American cities such as Atlanta, Dallas and Denver where there have developed new economies and political cultures that reflect success, growth and corporate confidence.

The cases demonstrate that city policy makers support regional policy innovation in economic development. Pittsburgh, Birmingham and Rotterdam need to remain competitive by building upon the successes of their most innovative enterprises. Global competition has led to the relocation of economic activity as corporations locate manufacturing on green field sites away from older cities. Policy makers therefore support the development of local labour skills that contribute to regional growth and they seek new ways to connect local communities to expanding regional economies.

The selected cities have active public–private partnerships crucial to improving competitiveness and leading economic development and urban regeneration. Corporate and public policy makers organise in policy networks, and they have joined partnerships that form part of the complex politics of urban regions. Public officials, politicians and business leaders strive to shape urban futures rather than simply respond to circumstances. Networks and partnerships coordinate groups to overcome urban decline, reduce unemployment and achieve economic growth. Partnerships have produced 'shared visions' that bring corporate and community interests together in a variety of programmes.

The selected cities have developed holistic policies to deal with acute social problems and the problems of disadvantaged communities. Pittsburgh, Birmingham and Rotterdam have socially distressed communities where groups experience discrimination and social exclusion. Manufacturing industries in Birmingham declined and deep social divisions led to urban riots and racial friction in the 1980s. As the big cities of the Netherlands became a refuge for people from East Europe and the former Dutch colonies, extreme right wing political groups in Rotterdam demanded the end to immigration as a solution to urban problems. In Pittsburgh, ethnic minorities have often been left behind in the regional economy and destined to live in poor and deprived neighbourhoods. These are concerns for policy makers who increasingly integrate social goals into urban regeneration programmes that 'holistically' foster joint working between different agencies and groups.

Research propositions

The above considerations ensured that there were common points of reference between the cities in the comparative study of economic development strategies. The main research questions and issues focus upon the relationships between regional policy strategies and partnerships set against the backdrop of the social and economic conditions mentioned above. Identifying the characteristics, trends, and paradoxes associated with public–private partnerships and policy networks produces the following research propositions that are at the heart of the present study.

Big city governments still seem to be politically important in spite of economic restructuring. Global competition and the regional restructuring of economies produce great variety in policy and organisation. Despite rapid change, big city governments have been remarkably resilient when confronted with global and regional restructuring, and the city councils in Pittsburgh, Birmingham and Rotterdam have all retained and developed important roles in developing new regional policy strategies. Market-led policies, privatisation, networking and partnership have forced city governments to reassess their policy interests, and the continued agglomeration of economic activity within wider regions (Barnes and Ledebur, 1998; Scott, 1998) has led cities to expand their regional involvements. Cities have therefore redefined their roles and relationships with corporate and nonprofit organisations by establishing networks regionally and locally.

Public policies are set within different, and profoundly changing, political cultures. Policy makers in each of the case study cities adopt national and culturally influenced views of the world. Rose (1991) therefore warns against making simple generalisations about the similarities of political systems because values and expectations can differ widely. Cultural differences are reflected in institutions and policies, and cities have political traditions that subtly produce differences of meaning in public policy. For example, Thompson, Ellis and Wildavsky (1990) detect political and social tensions arising from attempts to effect a transition from hierarchy in government to a more market-oriented approach. This gives rise in big cities in the USA, Britain and the Netherlands to different interpretations of the market model and contrasting ways of challenging old hierarchies and traditional policy priorities.

Bureaucrats commonly seek to restrain political competition to reduce the risks of policy making. Major economic and social transformations have disrupted old political certainties and increased the risks associated with policy. City governments and public agencies work with corporate and community leaders to achieve mutually beneficial strategies that reduce uncertainty, but partnerships can be risky, potentially unstable, and sometimes inefficient providers of services. Success depends upon order to provide conditions necessary for successful outcomes, so public officials and politicians try to 'restrain' political competition in networks to achieve shared visions and strategic control over policy. However, the policy process is highly political, and groups jostle for resources and attempt to influence policy outcomes. Community leaders seek access to politicians and public bureaus join with nonprofit organisations and companies, but groups compete with one another for public funds and political recognition. Partnership implies that groups recognise the advantages of coming together to achieve common goals, increase their resources and reduce conflicts. Networking is therefore crucial to the groups concerned, yet collaboration usually depends on delicate political settlements between partners that require them to play-down their differences and seek shared strategic vision.

There is no straightforward relationship between policy strategy and the structure of partnerships. The relationship between policy strategies and the regional organisation of policy networks and partnerships is complicated. Even apparently innovative policy strategies seem often simply to strengthen existing public organisations and institutions. For example, in Britain the former Conservative government often preferred to rely on established policy networks in regional policy. After 1997, the new Labour government proposed Regional Development Agencies to oversee urban and regional programmes, but even Labour's more far-reaching reforms adapted and rationalised existing programmes and administrative processes. Public–private partnerships have frequently developed independently in spite of grand visions, and strategies often follow institutional change. In all three cities studied, regional strategy therefore provides but one context for the organisational development of individual partnerships. While strategy is important in policy terms, any one type of strategy generally does not appear rigidly to directly produce any one particular type of partnership structure (see chapter two) although strategic commitments clearly do affect management processes and organisational reform. Partnership structures shape as the result of complex contingencies both in the internal environment and within government and they are influenced by political competition and the specific conditions prevailing in particular policy systems. They can thus be strongly influenced by attempts by governments to restrain and control networks.

'Fuzzy organisations' and holistic programmes are conducive to innovation, but top–down structures and policies develop even in the most innovative partnerships. The lack of a straightforward relationship between strategy and the organisational characteristics of economic development partnerships underscores the apparent lack of coherence within and between networks. Partnerships can be organisationally flexible and quick to adapt to changing market conditions, but structures vary widely and many partnerships fail to consolidate important core organisational competencies. Change in the external environment requires innovation in policy, the sharing of risks, and organisational responsiveness for which partnerships are well suited. However, collaboration frequently produces organisational hierarchy especially when public officials try to restrain political competition and control events.

Networking often limits the accountability of public agencies and local governments and it can hinder the empowerment of local community groups. While networking provides opportunities for groups to gain access, hierarchy in networks can reduce public accountability and sometimes lead to the exclusion of community organisations from the policy process. Different local conditions produce a variety of modes of participation and consultation. In the Netherlands, 'public sector' intervention involves structured consultation with interest groups arising from a consensus politics and the inclusion of groups. In Britain and the United States, groups often fail to influence policy because policy makers neglect community needs and demands.

The organisation of the book

The book is organised as follows:

Chapter 1. Culture and Change

City governments and public–private partnerships exist within different national cultural contexts that influence patterns of governance. Market policies in economic development have challenged public bureaucracies and encouraged administrative fragmentation. Cities have commonly 'reinvented' and reformed local public services, but national cultures influence political relationships and institutions.

Chapter 2. Strategy, Partnership and Contingency

The contingency model adopted in the book analyses policy networks and the relationship between strategies and partnership structures. It emphasises the organisational problems of networking and the attempts of policy makers to overcome these through managerial rationality and holistic polices. The model accounts for the emergence of a range of new structures and political conflicts within policy networks. The model presents economic development partnerships in the context of their external environments and highlights the factors that influence policy strategies and the organisational characteristics of partnerships.

Chapter 3. Strategy and the Competitive Environment

Regional strategies 'position' cities to take advantage of market conditions and reduce the risks of economic restructuring. Competitiveness depends upon innovation and the re-engineering of corporations to achieve dynamic economic performance. The chapter covers the different policies adopted in urban regions with reference to the difference between intuitive strategy and strategic planning.

Chapter 4. Strategy and Partnership in the Pittsburgh Region

The case studies of Pittsburgh, Birmingham and Rotterdam each start with a discussion of specific regional strategies and partnerships, followed by city and community strategies. The research in chapter four shows that strategy in Southwestern Pennsylvania aims to build regional competencies through strong business leadership, and this approach is also applied in local communities.

Chapter 5. Pittsburgh: Partnership and Community Empowerment

Large corporations have a long history of establishing nationally acclaimed urban regeneration initiatives in Pittsburgh. The revitalisation of the central business district has transformed the city through municipal and private effort, and neighbourhood groups work with business in innovative partnerships.

Chapter 6. Birmingham: Strategy and Partnership in the West Midlands

In the West Midlands region, the rich diversity of partnerships illustrates the growing importance of business involvement in strategy making. Public officials, corporate managers, and nonprofit organisations participate in a variety of regional initiatives. However, in spite of their potential flexibility, partnerships in the region commonly organise hierarchically.

Chapter 7. Birmingham: Partnership and Community

Local policy is pragmatic in Birmingham, being influenced by the need for the city to attract central government funds and inward investment. Policies preserve a strong role for local government action through extensive consultation with community organisations, and attempts to achieve more holistic approaches to policy are in line with the Labour government's attempt to extend joint working between agencies and groups. Holism in this setting is about collaboration and the projected 'seamless working' of initiatives linking different agencies and organisations.

Chapter 8. Rotterdam Scenarios

Strategy in Rotterdam focuses on the development of the city's surrounding metropolitan area and the competitiveness of the city as a World Port. Regional organisations have contributed to the development of strategic scenarios that identify a range of possible outcomes for the city.

Chapter 9. Rotterdam: Partnership and Community

Partnerships are vital to the success of urban revitalisation in Rotterdam. Programmes incorporate community interests in urban planning with a strong emphasis on social integration and renewal.

Chapter 10. Evaluation

No magic success formula or winning strategy for economic development exists. Each city has a distinctive process of change that requires sensitive evaluation taking account of local circumstances.

Economic development and urban regeneration

Economic development and urban regeneration are presented in the book as together defining a policy domain in the way that education and health constitute broad policy domains. Boin and 't Hart (1998) refer to policy sectors such as immigration, social security, and agriculture and fisheries that are susceptible to change and crises in fast changing political and economic environments. They define a policy sector is an 'institutional field of actors, rules and practices associated with state efforts to address a particular category of social issues and problems' (Boin and 't Hart, 1998, p. 4). At the centre of a policy sector are key government decision makers but important within the sector are 'also the semi-public and private organisations that may be involved in the implementation of policy. In addition, there is often a cluster of clienteles, pressure groups, lobbyists, and other government agencies that have a stake in sectoral policy and try to influence its scope and content' (Boin and 't Hart, 1998, p. 4). In this book, a policy domain is even more broadly defined as clustering the public and private organisations that have interests in policy issues associated with the domain. The domain concept is like that used by Knoke, Pappi, Broadbent, and Tsujinaka (1996) which subsumes policy sectors while allowing for an even wider variety of ambiguities and organisational and network configurations. Associated with a domain, there are different partnerships of groups and organisations and various informal networks. Government and business work with and within these clusters of groups in various ways to effect policy outcomes and sometimes corporate or business interests can substitute for government in defining the policy agenda, as in the USA where there are strong business leadership groups. The policy domain here is not exclusive in the way that a narrowly defined 'policy community' is (Marsh and Rhodes, 1992), although within domains there are hierarchies of groups and influential policy makers. The policy domain is wide and fuzzy-edged and overlapped with other domains, but it has a substantive policy focus such as health, education or economic development that brings groups together.

Economic development involves stimulating economic growth and competitiveness in regions and cities through the attraction of inward investment, company formation, and job creation activities (European Commission, 1998). Economic development and urban regeneration coexist in the same policy domain because of the close interrelationships between the corporate, nonprofit, and community groups with interests in economic growth and regeneration. Public policies in the 1990s link economic development and urban regeneration to establish connections between local communities and regional economies. Urban regeneration has many meanings according to different national and organisational preferences, but here it is about the physical and social revitalisation of cities which increasingly impinge upon economic competitiveness and the development of the workforce skills of regional economies. This inevitably puts urban regeneration on the agendas of organisations involved in economic development, and it implies that policies are subject to market risks arising from changing conditions.

The focus in this book is on partnerships, although many collaborative relationships between government departments and public agencies are not formally partnership arrangements. For instance, the British government has encouraged closer joint working between public agencies and government departments as part of a drive toward 'joined-up government'. Inevitably, in economic development and urban regeneration the distinctions between interdepartmental collaboration and partnerships will blur as public and private organisations work together in networks and development partnerships. Joined-up government assumes that public–private linkages will increase and that seamless working between organisations will be commonplace.

1
Culture and Change

Big city governments increasingly make policy strategies to deal with the combined challenges of global competition and the restructuring of regional economies. In the USA, Britain, and the Netherlands, politicians have advocated freer markets and public–private partnerships as alternatives to old-style state intervention. However, there exist deeply rooted and culturally defined social processes that influence the shape of the political structures and interrelationships that develop. This chapter therefore concentrates upon some fundamental changes in the wider social and economic environment influencing public policy and their implications for organisational hierarchies and modern governance. The policy domain of economic development and urban regeneration (see Introduction) is affected by these changes, and this chapter provides the context within which economic development issues are dealt with by governments and private corporations.

The context of change

Clearly, important changes are taking place that influence the governance of cities and communities. Rhodes argues that modern 'governance' is changing 'the meaning of government,' as part of 'a new process of governing' involving the proliferation of 'self-organizing, inter-organizational networks' (Rhodes, 1997, p. 15). One aspect of this change was the political tide that made popular the appeal of politicians such as Reagan and Thatcher in the 1980s who advocated new approaches to managing and delivering services. A wave of 'New Public Management' in many developed democratic countries, including Britain and the Netherlands, gained the support of public administrators who in their varied and often eclectic ways sought diversity in the way that policies could be implemented (Dunleavy, 1991). They strove for greater managerial efficiency, public–private partnership and management autonomy (Skelcher, 1998). In the USA, Osborne and Gaebler (1992) shared the concerns of advocates of European New Public Management (Rhodes, 1997), but Osborne and Gaebler had a special concern with markets and competitiveness and argued that centralised government was poor at delivering, or 'rowing' services. New approaches to management and networked implementation were therefore necessary to meet the challenges of an increasingly globalised economy because they were about flexibility, responsiveness to change and economic well being. Goetz (1993) refers to interrelated global processes that redefine urban governance through economic and spatial reorganisation that, combined with political reforms, produces a 'new localism'. The new localism involves the restructuring of public organisations and attempts by policy makers to rationalise relationships between different levels of government to address global market challenges (Goetz, 1993, p. 202). International comparisons of public policies reveal local governments balancing different political interests (Mackie and Marsh, 1995) as there emerge new trends, political contexts and changing cultures. In the European Union, the drive for regional cohesion, sustainability, cross-border partnerships, and new linkages between cities and regions is forcing dramatic changes in the ways governments work and in attitudes about national identities and interests.

Clark (1994) identifies a New Political Culture that challenges old government hierarchies and strengthens market individualism. Using data from the Fiscal Austerity and Urban Innovation Project, he shows changes in the attitudes of citizens and policy makers in countries where governments strive for global competitiveness. Clark and Hoffmann-Martinot (1998) show that the New Political Culture blurs old left–right political distinctions and produces new political groups. Under the old class politics, social concerns were subordinate to economic issues, but this relationship changes as 'correlations between social and fiscal liberalism decline' (Clark, 1994). Greater emphasis on issues such as gender and the environment develops as people become more concerned about lifestyles. This accompanies declining support for big government and more attention to service efficiency, although new concerns coexist with established patterns of public welfare and the defence by some groups of old structures.

Meyer and Scott (1992) describe the changing external environments of public institutions as 'cultural systems' influenced by change internationally (Meyer and Scott, 1992, p. 1). They argue that the customs and ideologies of public officials produce distinctive institutional characteristics in public organisations and that laws, customs and ideologies pass to governments as 'procedures flow from organization to organization, sector to sector, and even country to country' (Meyer and Scott, 1992, pp. 1–2). The ideology and culture of a particular public institution therefore bring together different values, organisational practices and management philosophies. Meyer and Scott argue that local policy makers adopt organisational structures by borrowing from central government and incorporating external policy experiences 'into their own systems' often extensively importing 'tools, techniques, skills, know-how' (Scott, 1992, p. 16) from outside. Peters (1988) also shows that political systems do not simply reflect national conditions, but global change also influences them. The 'universe of organizations' (Peters, 1988, p. 18) structures according to spatial and organisational complexity, and public bodies undergo administrative 'succession' with new organisations replacing old ones in uncertain political environments.

Cultural theory

Thompson, Ellis, and Wildavsky (1990) associate public institutional change with change in society through a grid–group typology (see Figure 1.1). Grid and group are 'ordering principles in constituting ways of life' (Ostrom and Ostrom, 1997, p. 85) with a continuum that measures how a 'grid' of social variables affects individuals. High grid describes a social context in which highly regulated individuals relate to external institutions. High grid means that individuals do not negotiate 'their own relationships with others' (Thompson and Ellis, 1997, p. 2). High grid therefore implies a low degree of negotiation binding individuals to 'externally imposed prescriptions' (Thompson *et al.*, 1990, p. 5). In a developed market economy with little government intervention and decentralised private enterprises, individual negotiation is normally high. In Figure 1.1, 'group' shows that individuals are 'incorporated into bounded units' (Thompson *et al.*, 1990, p. 5). Therefore, the incorporation of individuals into political institutions reduces the individual choices that

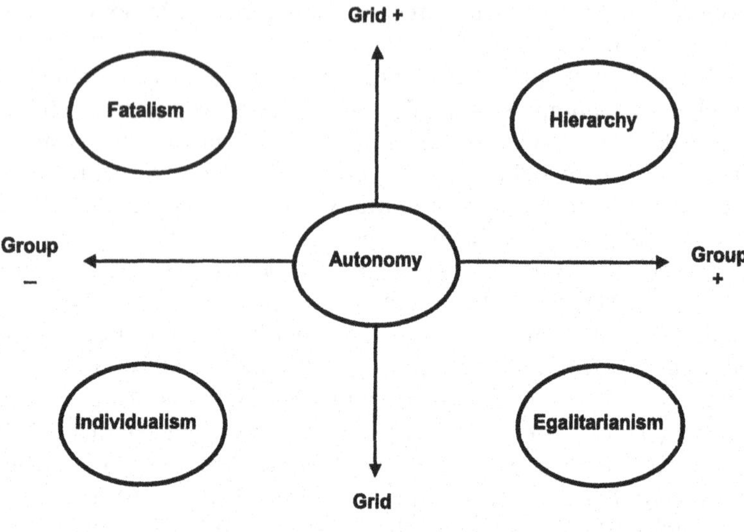

Figure 1.1 Grid-Group Typology

Source: Adapted from Thompson, Ellis and Wildavsky, 1990, p. 8. Copyright (c) 1990 by Michael Thompson, Richard Ellis and Aaron Wildavsky. Reprinted by permission of Westview Press, a member of Perseus Books, LLC.

they enjoy. In the 1990s, this group association declined as labour union membership fell and community organisations took up single issues rather than defending old collectivist ideological positions and political parties.

Cultural biases and social relations form 'ways of life' that influence values and policies. Thompson *et al.* (1990) depict hierarchy (Figure 1.1) as one possible way of life where external rules, regulations and organisations constrain individuals. In contrast, egalitarianism allows the participation of different groups through public consultation and accessible institutions. Thompson *et al.* (1990) show that conservative politicians in the 1980s encouraged market individualism to counter the inefficiencies of big government, but such policies often unintentionally produced fatalism, depoliticisation and welfare dependency. Fatalism reduced the influence of collectivist organisations, like the labour unions, but enabled governments to carry out policies often against the views of organised interests.

For Thompson *et al.* (1990), cultural bias underlies shared values and beliefs in political systems. Interpersonal relations produce social relations that, with cultural bias, produce particular ways of life. Social relations and cultural biases mutually reinforce conceptions of politics and government that strengthen social and cultural values. A way of life, however, does not remain in equilibrium because the constant interplay between changing values and social relations induces change. The vulnerability of a particular way of life depends on the strength of the prevailing relationship between cultural biases and social relations so that a way of life is sustainable only if 'it inculcates in its constituent individuals the cultural bias that justifies it' (Thompson *et al.*, 1990, p. 2). Predominant values in a way of life are stabilising, and individuals that support the way of life are often aware of what behaviour is functional to its maintenance. Change occurs when 'successive events intervene in such a manner as to prevent a way of life from delivering on the expectations it has generated, thereby prompting individuals to seek more promising alternatives' (Thompson *et al.*, 1990, pp. 3–4). For example, in the 1980s public opinion in Western Europe and the USA shifted to favour greater market individualism. Thompson *et al.* (1990) argue that the change involved a variety of strains and conflicts as institutions and political groups competed. This set in motion contradictory processes that were often left unresolved. For example, British Prime Minister Thatcher wanted to limit the role of the state, but the proliferation of public organisations and centralised government created new public and private organisational hierarchies. The group–grid approach shows how contradictory forces mix up public and market outcomes. It shows how the fatalism of the unemployed can coexist with the dynamic successes of knowledge based industries and the egalitarian demands of community organisations.

Culture and organisation

Using the grid–group approach, Coyle (1997) claims that a cultural theory of organisations can be consistent with mainstream organisational theories. For Coyle, it is better to regard different ways of life as generating 'tendencies or pressures rather than operational forms' (Coyle, 1997, p. 64) since many hybrid kinds of organisation can develop. Both public and private organisations can be hierarchical although they relate to the market in distinctive ways. As Coyle (1997) argues, a libertarian way of life based on pure market individualism would imply bargaining and coalition building between individuals with a decreasing reliance on formal organisation. A pure market would bring new ways of organising affairs with individuals working within the market and becoming less dependent on the external organisational hierarchies that previously regulated their lives. Coyle asserts that though 'it is fair to say that a low group, low grid culture will generally be compatible with a relatively autonomous private sphere, it does not follow that governmental control or administration will be illegitimate. Conversely, a hierarchical culture may have a minimal state, even though it is compatible with expansive bureaucracy' (Coyle, 1997, p. 70). Coyle views classic 1930s American progressive era bureaucracies as high grid, but while these were appropriate given the conditions of the time (Chisholm, 1997), more recent public bureaucracies have had to deal with new complexities. Networked organisations became more prevalent because they were better suited to unpredictable market conditions, but the 'cumbersome private hierarchies' (Coyle, 1997, p. 67) that frequently resulted were as bound by routines and rules as bureaucratic governments. The networks and partnerships in economic development therefore are often organisationally hierarchical because policy makers have tried to control change and bring order to fragmented political systems. Real world situations frequently involve moves 'outward and upward from markets' on the grid–group diagram from 'informal and ever changing networks to increasingly formal and rigid relations' (Coyle, 1997, p. 70).

The business culture in the USA

American political culture has historically emphasised the virtues of enterprise and pluralism against state hierarchy and interference. According to traditional accounts, citizens enjoy democratic freedoms and the constitution protects individuals. Americans have a federal system of government that reinforces these freedoms. Subnational government consists of various authorities with state, county and local governments adhering to democratic values that allow individuals to express their demands. However, despite the democratic commitment, such a description of government operating at the federal, state and local levels is no longer adequate. In the 1950s and 1960s, some writers modified this interpretation by reference to 'iron triangles' of influential interests that competed for resources and influence. Heclo (1978) questioned this view claiming that fragmented government comprised of issue networks that were unstable and likely to lead to policy drift and uncertainty. Nevertheless, Kincaid (1994) describes the modern government in the USA as a paradoxical combination of federal influence over policy but with a strong role for the states. He shows that the federal government has substantially influenced domestic law, and even in the 1980s the government assumed greater regulatory authority. Nevertheless, state and local governments levied taxes and consolidated their 'leading fiscal and administrative position in domestic policy' (Kincaid, 1994, p. 200). Kincaid shows that state and local government revenues grew faster than those of the federal government during the 1980s as subnational governments expanded their activities domestically and internationally (Kincaid, 1994, pp. 200–1). State and local governments successfully competed for inward international investment, and developed links with the corporate private sector in economic development, trade and commerce.

Peele, Bailey, Cain, and Peters (1994) describe the fragmentation of the modern political system in the USA and the conflicts within and between federal government departments and public agencies. They describe a complex 'multilayered federal system' that displays diversity in institutional design and culture (Peele *et al.*, 1994, p. 7). Diversity is important in any definition of modern governance as its recognition challenges a coverall description of 'networked governance'. Peters (1998) is cautious about describing the outcome of multi-layering and fragmentation in terms of networking since fragmentation does not necessarily by itself produce inter-organisational collaboration. He doubts the universality of networking by contending that the American system may be fragmented, but public agencies and local governments do not always work together. Indeed, agencies and government departments have occasionally operated with little sharing of resources and few interdependencies (Perrow, 1986).

Reservations about networking are valid, but interagency collaboration and the formation of policy networks involving consumers and interest groups are increasingly producing important new linkages between different interests. This is especially so in economic development and urban regeneration where local government policy makers work outside traditional government frameworks and in public–private partnerships (Walzer and Jacobs, 1998). The restructuring of markets encourages state and local governments to develop new policy initiatives to create jobs and regenerate communities. State and local governments work in a commercially competitive world, so they create innovative networks and partnerships to stay ahead. The combination of fragmented government with a dynamic private sector thus provides conditions that are conducive to local organisational variety and complexity.

Fuchs (1992), studies complexity in fiscal matters in the USA, presenting a picture of government with overlapping jurisdictions, special districts, authorities, commissions and county governments. In some cases, state governments have taken over responsibility for services while functional consolidation shifts 'the legal burden of a function to a higher level of government' depending on 'state constitutional or statutory authority'. The state legislatures 'play the critical role in changing the relationship between governmental units either mandating or facilitating functional transfers' (Fuchs, 1992, p. 192). Fuchs emphasises the constraints that the states place on cities and the political tensions between them that result. Conflicts develop over expenditure mandates, levels of taxation and municipal public debts. Critical in relationships between federal government and the cities is the allocation of federal aid, with frequent conflicts between politicians at the federal and local levels over funding. Disputes have often centered round the role of federal government, the interests of cities, and social benefits for the urban poor. Such conflicts involve the interests of city governments defended by mayors, the attitudes of public officials and the commitments of members of overlapping government bureaus and agencies (Fuchs, 1992, p. 180). Fuchs argues that the role of state government 'in formulating local fiscal policy limits the authority and discretion of cities over their own revenue and expenditure policies.' The relationship is based on a legal doctrine that gives the state 'supremacy over its political subdivisions' (Fuchs, 1992, p. 180). Fuchs explains how the precise fiscal relationship between federal government and cities varies with the older northeastern cities tending to limit the development of overlapping jurisdictions to deliver services and concentrating upon the expansion of existing functional units. In the Midwest, cities display a greater propensity to develop 'regional' authorities and complicated patterns of functional responsibility.

The business culture in Pittsburgh

This administrative fragmentation in Pittsburgh and other cities has been influenced by the historical relationship between, federal, state, and local governments and an expanding market capitalism. Monkkonen (1995) studies local financial dependency in American cities in the late nineteenth and early twentieth centuries and the emergence of the new financial climate following the 1930s Great

Depression. For Monkkonen, cities were politically weak in the nineteenth century, but they performed important economic functions as powerful business elites looked to city governments to support projects that contributed to civic development and as cities borrowed money to fund infrastructure projects. Monkkonen (1995, p. 19) argues that this produced 'a buccaneering sense of political enterprise' in the cities that lasted up to the Great Depression. Sbragia (1996) also adopts a long historical perspective that portrays entrepreneurial cities within a constantly changing federal system. Local government and private sector collaboration arose from federal intervention in the economy that expanded the role of the public sector. Sbragia argues that the 'evolution of the American private sector was intertwined with public power, frequently exercised at the state and local level, much more directly than was the British private sector' (Sbragia, 1996, p. 19). The American model of business leadership contrasted with that in Britain, where corporations were historically reluctant to get involved in local economic development. Sbragia consequently emphasises the importance of American state and local governments in an evolving relationship between the public and private sectors where the two became closely intertwined in local governance.

Despite the problems confronted by city governments following urban economic decline and suburban growth in the 1960s and 1970s, cities expanded their roles in urban regeneration and local economic development as they developed new roles working with the corporate private sector. In the 1980s, local governments therefore were remarkably resilient because they adapted well to changing conditions. Morgan and England (1996) argue that during the 1980s these new public–private relationships and the transformation of local government produced many different strategies to deal with economic restructuring. Pittsburgh illustrates the wide variety of public–private sector initiatives that evolved in a strongly entrepreneurial environment where the corporate private sector provided both regional and civic leadership. Political leadership through the Mayor, Deputy Mayor and small nine member city council was highly effective in organising policy by bringing together business, political interests and public officials. The influence of the private sector continued in the 1990s as Democratic Mayor Tom Murphy developed policies with companies in new high technology and service industries, and as he promoted the city as a good place to invest. Pittsburgh demonstrates how cities can develop a strategic regional role while maintaining a comparatively small municipal bureaucracy that supports effective business leadership. The city closely monitored public funding and the running of council departments to ensure increased efficiency, and this helped the city council control a budget deficit, albeit through forced municipal job reductions and severe strains on local services. The New Political Culture meant that big bureaucracy therefore declined as a symbol of political success or power. Rather, it was rationalised management combined with effective local and regional networking that enhanced the reputations of political leaders and their cities.

Hierarchy and market in Britain

As in the USA, British central and local governments carried out managerial and structural changes during the 1980s and 1990s, but there was a different relationship between the public and private sectors in economic development. The Thatcher years led to important changes to the classic model of British unitary government with its centralisation and hierarchical institutions. City governments funded local services and civic projects, but while they worked with the private sector, American-style business leadership was underdeveloped.

For Prime Minister Thatcher, state interference had failed, and the 'enterprise culture' would emulate the best aspects of the American market economy. Conservative governments between 1979 and 1997 privatised industries and contracted out municipal services, and they introduced new management techniques into the public sector. To combat the influence of left wing local governments, Thatcher abolished big metropolitan county authorities in 1986. For Birmingham, this removed the strategic metropolitan authority that covered the city. Yet Thatcher's policies generated both centralisation and fragmentation. The Conservatives encouraged business leaders to join local and regional public–private partnerships, but the government also transferred powers away from elected national and local governments to nonelected bodies that were less likely to be influenced by left wing Labour local authorities. The powers of local governments were reduced, but there were stronger central controls over important areas such as education and urban policy.

Birmingham in a new environment

Gray (1994) examined the expansion of nonelected subnational government (SNG) and the public organisations that existed 'away from Whitehall and Westminster' (Gray, 1994, p. 1). He maintained that 'subnational government was so complicated that it was 'difficult to monitor everything that takes place under such a system' (Gray, 1994, p. 1). The range of organisations included local governments, health service bodies, the regional and local offices of central departments and decentralised agencies.

In spite of this complexity, it was often difficult to get business leaders to work with local governments. Before Thatcher, direct business involvement in civic affairs was on a modest scale because corporate leaders concentrated on industrial issues, leaving local politicians to deal with local affairs. Ironically, Thatcher's political assault against Labour controlled city councils under the sway of the 'new urban left' (Stewart and Stoker, 1994) actually led Labour local authorities to work more closely with business. Thatcher substantially reduced the ability of cities to initiate publicly funded projects, but Labour councils in Birmingham, Manchester, Liverpool, Glasgow and elsewhere worked with the private sector to devise new growth strategies using partnerships and multiple sources of funding. This opened opportunities even though cities were constrained by central funding cuts, expenditure controls and the weakening of local government in the provision public services. Stewart and Stoker (1995) argue that the 'institutional map' of local government under Thatcher produced a system of local governance whereby local authorities sacrificed their autonomy in traditional areas but began to work extensively with a wider range of non-government organisations. The system had 'become increasingly differentiated as new agencies and organisations have been given responsibilities which previously belonged to local authorities or as existing institutions have been removed from the control of local authorities and health authorities' (Stewart and Stoker, 1995, p. 194). Local government between 1979 and 1997 therefore became more differentiated, complex, and varied as new interactions developed linking central government, local authorities, nonprofit groups, intergovernmental boards and business organisations.

Despite the comparatively weak business leadership culture, business did gradually become more involved and proactive in urban regeneration. In the early 1980s, Business in the Community promoted corporate social responsibility in local communities, taking the American business leadership model as a guide. Business in the Community expanded its activities in cities throughout Britain working with local authorities and nonprofit organisations. Business leadership groups were established and public–private partnerships benefitted from the involvement of local business leaders and national corporate donations and support. Also, in the 1990s, the British Urban Regeneration Association supported business involvement in urban and regeneration by encouraging local authorities, property development companies and voluntary groups to work together in improving urban areas. The Urban Villages Forum also played a role in linking companies to local authorities and communities through a range of community planning and environmental improvement initiatives. In Birmingham, these organisations were active alongside business groups based in the city and the Birmingham Chamber of Commerce. The Chamber of Commerce had an active interest in urban regeneration and local economic development with close links with the city council and development agencies.

With the election of Tony Blair's 'New Labour' government in 1997, promarket policies and the complex labyrinth of local networks and partnerships remained, but there were new policies for the regions and local government. Labour was committed to boosting regional competitiveness and strengthening regional economies through major constitutional and economic reforms. Local governments, starting with London, were to have democratically elected mayors similar to those in other European Union countries. In Birmingham, the Labour party controlled city council placed partnerships high on the urban policy agenda. This confirmed what Hendriks (1996) regarded as a weak egalitarianism in Birmingham, and he also detected a traditionally strong reliance on hierarchy with large local government departments, centralised programmes and a 117 member council. But, during the 1980s, local government hierarchies were substantially modified as the city developed community programmes and adopted a broader international market orientation. The city also maintained close contact with business groups and encouraged new public–private sector partnerships. Under the Blair government, the ruling Labour group on the council, under the leadership of Theresa Stewart and later Albert Bore, could therefore extol the virtues of business collaboration and local community empowerment in local services. 'Best Value' and holistic policies would produce more efficient and citizen-responsive services. Theresa

Stewart and Albert Bore thus enthusiastically supported Blair's commitment to modern local government and stronger leadership in local government to enhance the city's competitive economy.

A Dutch third way?

The situation in the Netherlands is complicated in cultural theory terms, but here again there is a desire to foster public–private sector cooperation by breaking-down old public hierarchies. Nevertheless, social and economic change simultaneously challenges and draws inspiration from traditional values, institutions and policies. Admired by Tony Blair, the Dutch have been attributed with developing a 'third way' in social and economic development, and they have produced some unique cultural outcomes providing a contrast to old-style European social democracy and American and Thatcherite neoliberal market models. The Dutch have reformed the labour market and achieved a flexible market economy, yet they have also managed to retain a social safety net with generous welfare benefits. The Netherlands avoided a Thatcher-style conflict with the labour unions and local government and escaped the kind of labour unrest experienced in France and Germany in the 1990s. However, as Giddens (1998) argues, the Dutch alternative to neoliberalism and public interventionist social democracy has been less successful in providing a third way than often is claimed. For Giddens, the 'Dutch model', while combining a deregulated economy with a social safety net, has failed to redistribute resources towards the development of human resources. Instead, unproductive social provisions serve to reinforce the welfare dependency of the unemployed and other claimants.

Nevertheless, part of the explanation for the comparative attractiveness of the Dutch approach to outsiders has been attributed to the consensual nature of national politics in the Netherlands. Lijphart's (1975) account of the 'politics of accommodation' provided a classic view of how political elites bridged social cleavages in Dutch society through their shared desire to arrive at a political consensus. The historical record of religious and cultural diversity was strengthened by the 'pillarised' institutions that brought together key social groups and political parties and produced an orderliness in politics, institutions and economic affairs. The pillars of society provided the nation with a conflict-reducing form of politics that ensured support both for public services and pro-market policies.

In the Netherlands in the 1990s, contemporary politics involved a redefinition of the consensual traditions of the political system within the context of the competitive global market, an ageing population, and greater racial and ethnic diversity. These forces led to a reassessment of the 'polder model' of consensus that was conducive to managerial efficiency, but which often produced political inertia with politicians 'wading through treacle' before workable compromises could be made (Organization for Economic Cooperation and Development [OECD], 1998).

For Rosenthal and Roborgh, adaptation to new conditions created an 'uneasiness' that tempered the drive for change. The Dutch emphasis on consensus and coalition led to government with strong corporatist and hierarchically organised characteristics (Andeweg and Irwin, 1993; Rosenthal and Roborgh, 1995) that shaped key policies and government institutions. The incorporation of the political parties and other interests into the political system reduced conflict and cleavage to what Rosenthal and Roborgh (1995, p. 349) describe as 'mutually accepted limits'. Andeweg and Irwin (1993) refer to the conflict regulating mechanisms and the proportional representation of interests that ensured continuity in policy, but this was accompanied by the depoliticisation of important social issues (Rosenthal and Roborgh, 1995). In the 1960s, depillarisation challenged traditional political elites through decentralisation and democratisation. However, tradition coexisted with the forces of change (Rosenthal and Roborgh, 1995) and the trend towards administrative fragmentation. In the 1990s, central and municipal governments experienced accelerated reform in response to economic restructuring, international competition, and political and social conditions, but the tensions between change and continuity permeated the administrative reform process frequently creating conflict in place of the consensus.

Changes in the economy and government were confirmed with the privatisation and semi-privatisation of state enterprises. In 1998, 15.6 per cent of the working population in the Netherlands were employed in the public sector (OECD, 1998) despite the efforts of governments to reduce the size of government. However, privatisation in the Netherlands continued, but unlike in Britain it has been regarded by successive governments as necessary to achieve administrative efficiency as opposed to ideologically driven political objectives. The government of Prime Minister Lubbers after 1982 shifted the emphasis to 'more market and less government' (Haffner and Berden, 1998) and initiated an evaluation of the 'public enterprise sector'. This sector consisted of a variety of bodies and activities that included government enterprises and services and enterprises with government shareholdings. A process of corporatisation under private law created a range of autonomous enterprises that had limited state involvement and which operated commercially. Other enterprises, under a policy of functional decentralisation, were kept within the public sector but operated by autonomous bodies, as in the case of social security administration.

Decentralisation

In local government, municipalities have been affected by decentralisation and extensive management innovation. Local and provincial governments obtain ninety per cent of their financial resources from central government grants (OECD, 1998), and the government has sought to better distribute these resources and provide local governments with greater influence over how they are used. Andeweg and Irwin (1993) describe autonomy and co-government in modern Dutch politics where constitutionally, municipal governments enjoy autonomy, but they also have responsibilities for implementing national legislation. The twelve elected provincial authorities also play a role, and the country divides for statistical and planning purposes into major regions (north, south, east and west). Forty Central Bureau of Statistics 'regions' provide a further breakdown consistent with European Union regional classifications. Andeweg and Irwin (1993) describe the relationship between the provincial and municipal levels as strained despite provincial consultative and advisory committees and improved municipal and regional cooperation. Municipal governments frequently regard provincial authorities as intrusive, and the larger cities, such as Rotterdam, The Hague, Amsterdam and Utrecht, have campaigned for new regional authorities. Andeweg and Irwin refer to the dysfunctional size of the provincial authorities that are small compared to subcentral regional authorities in other European Union countries.

The Netherlands Scientific Council for Government Policy (NSCGP, 1990) emphasised the responsiveness of Dutch institutions to changes in the global economy that influenced the scale and depth of institutional adaptation in the 1980s. Competitive pressures encouraged the Dutch to expand the role of local government and encourage institutional innovation in economic policy and enhancing economic competitiveness by strengthening local infrastructures and expanding high technology growth industries. Collaboration between state, private and nonprofit organisations was crucial, and it involved city councils in new multi-agency relationships. Central government, public and nonprofit organisations developed policies suited to local circumstances and devolved responsibilities to the local level (NSCGP, 1990, p. 22). The two major structural shifts influencing intergovernmental relations therefore were less reliance on government and the decentralisation of decision making to local authorities. The scientific council found that the centralisation in the Netherlands created imbalances with the public sector focusing on a narrow range of social needs. This produced a system in the 1970s that lacked the dynamic economic culture of American cities where local government freedom enabled cities to be more adaptive and innovative when working with business. In the 1980s, Dutch local authorities had more influence over centrally funded programmes and could spend money according to local priorities. Van der Loos (1993) claimed that this enhanced city governments as foci for social and economic development and Torbijn (1993) found a growing emphasis on infrastructure, business development, social renewal and integrated policies. Public–private sector partnerships and greater decentralisation in these areas meant that no 'single actor' (Sudarskis and Edwards, 1993) could so easily dominate policy.

The changed environment in local government led big cities to reassess their roles in the Dutch national economy. The government's policy for 1991–4, 'Regions without Frontiers', concentrated on international economic competitiveness and the need for policies to improve the environment for business at the regional and local levels. However, city governments were not always strongly committed to business. In Amsterdam, The Hague, Rotterdam and Utrecht progress was slow up to the end of the 1980s, but cities had to back business because they could not 'stand alone but must find their own culture and relationships in the business sector' (NSCGP, 1990, p. 77). Moreover, cities were encouraged by central government to adopt more strategic, results-oriented approaches to policies that would better equip them to deal with global market change and the social problems associated with changing local demographics. The transition away from public hierarchy therefore was not straightforward or easy, and local government in the Netherlands therefore combined both elements of centralism with market change and adaptation. Duffy (1998), referring to Bianchini and Parkinson (1993), therefore characterises Rotterdam's government in the early 1990s as technocratic-elitist working through municipal departments and agencies to implement the policies that political leaders want, often without direct public accountability. Indeed, as Eldersveld, Stromberg, and Derksen (1995) show, Dutch local political elites have less contacts with local interest groups than elites in the USA. However, at the end of the 1990s, while this description had some currency, there was a significant shift in favour of public–private action, local consultation, regional collaboration and a wider global policy perspective. These suggest a more appropriate characterisation of Rotterdam as a technocratic–consensual politics within a market environment that modifies traditional government hierarchies. Nevertheless, this is a culture where corporate and political party elites are strong and where informal networking is crucial to the work of local government. Eldersveld *et al.* (1995, p. 161) refer to the importance of business pressure groups in the Netherlands and their connections to top decision takers and central government departments. In this context, the technocratic consensus characterisation allows for the articulation by business groups and political elites of market policies despite the relative underdevelopment of the kind of partnering that is common in the USA. In that sense, American business groups tend to 'surface' more often in partnerships whereas in Rotterdam corporate interests often influence the local policy agenda more discreetly.

Rotterdam: adapting to the market

The importance of political networking is illustrated well in local government where central government appoints city mayors. In Rotterdam, as elsewhere, the mayor should not be regarded as a government placeman or an intractable defender of big government. The mayors are party members, but they are impartial chairs of their councils and, as a result, they have no council vote. This tends to depoliticise mayors, but in the modern world mayors are expanding their political horizons and, in the case of Rotterdam, getting involved in national, regional and international policy networks. Former Rotterdam Mayor, Bram Peper, performed an active role promoting the city and contributing to policy debates along with the local political parties, business and communities. Peper, a member of the Labour party (PvdA), in 1998 joined Prime Minister Wim Kok's coalition government of Labour, pro-market Liberals (VVD), and Democrats (D66). During his period as Mayor of Rotterdam from 1982, Peper attended to maintaining the high profile of the city through influential networks such as Eurocities of which he was President prior to his rise to national office as Minister for Internal Affairs.

As Mayor, Peper presided over Rotterdam city council with its forty-five councilmen, and he chaired an Aldermanic Board selected from council members. Following local elections and Peper's departure in 1998, the board worked as a multiparty coalition that included the Labour Party Deputy Mayor, two other Labour members, two market Liberals, one Green and one Christian Democrat. The board had oversight over municipal core 'sectors', companies, and professional advisory and consultancy groups. The sectors included Social and Economic Affairs, Social and Cultural Affairs, Housing and Infrastructure, and Policing and Security. Directors were responsible for the sectors and for relationships between them and the municipal companies. The directors worked with public servants organised into specialist teams and maintained close contacts with central government ministries. Central funds were important, but the city had considerable discretion over spending money on innovative programmes and initiatives. One example was within the Social and Economic Affairs sector spanning social, economic, investment, urban regeneration and port issues. Policies also related to information technology, economic affairs and social policy, each with divisional heads and project groups that advised senior officials. Public officials argued that city services had reduced centralisation and inefficiency through the introduction of business methods. Under these circumstances, municipal sectoral directors, company directors and city aldermen all contributed to the change agenda, but Rotterdam's influential economic elites also influenced policy especially through growth-driven economic development initiatives and their investments in Rotterdam's World Port. The local chamber of commerce provided an important link between the political and economic elites driving growth, and individual businesses influenced the city council through personal networks, the Mayor and public officials. Companies with interests in the regional economy, including Royal Dutch-Shell, could access the Mayor and aldermen while those in commercial property development had influence in informal networks and emerging partnerships.

New ways of working

How do all the changes mentioned above affect the ability of regional cities to control their own affairs?

The New Political Culture has altered relationships between local governments and business, but cities remain as important political players despite their economic weaknesses and the constraints imposed upon them by central governments. Certainly, some cities have come under economic pressure and have declined in political influence. For example, Tiebout (1956) maintained that urban politicians in the USA could do little to counter decline when, ultimately, it was the market that sorted out the complex economic decisions of voters and businesses. Peterson (1981) argued that cities were therefore highly susceptible to market forces that placed 'limits' upon policy choices. Later, others have assumed that big city governments are powerless to resist globalisation, and Ohmae (1990, 1995) refers to the end of the nation state and the rise of global regions in a technologically 'borderless world'. Some writers maintain that America's big cities are so vulnerable to global economic forces that city governments have little influence in the global market (Kantor, 1995; Sassen, 1995), and that the new dynamic culture and workings of the market speed their political decline (Schockman, 1996). Urban political coalitions are said to have weakened as economic activity locates at 'edge city' (Garreau, 1991). Urban regimes, as coalitions of community and business interests, find it difficult effectively to respond to changing market conditions as governments fragment into networks that politicians cannot control (Stone and Sanders, 1987). Similarly, 'hyperpluralism' produces diverse interests that contend for influence over urban policy, while wealth and political power moves to the suburbs and new growth poles (Thomas and Savitch, 1991).

However, these 'weak government' arguments neglect the changed role of cities and the location of cities in new national, regional, and local networks and partnerships. The reconfiguration of local networks was evident in the 1980s when King (1987) criticised Castells (1977), Smith (1979) and Katznelson (1981) for regarding urban policy makers as conditioned by forces that are simply beyond their control. Gurr and King (1987) argued that local governments could pursue their interests precisely because they worked more in conjunction with private, voluntary, and public sector groups. Swanstrom (1988) claimed that local governments enjoyed substantial discretion over policy and could favour poor communities without setting off investor migration from high tax areas. Big city governments could defend their interests independently of central government, and they could control their own economic resources (Gurr and King, 1987), even acting independently from professional groups that were 'capable of nationalizing policy irrespective of a locality's distinctive character' (King and Pierre, 1990, p. 2). These forces have been strengthened in the 1990s, and in spite of the loss of influence of many cities over their traditionally strong public provisions, city policy makers enjoy substantial political influence within public–private hierarchies that work aside from traditional local government structures.

Change produces many new opportunities for urban policy makers as well as producing problems for city government. Stone therefore advocates a reassessment of American urban politics, arguing that 'many practices' associated with the political machine 'linger on' through patronage and political favours (Stone, 1996, p. 450). Eldersveld *et al.* (1995) argue that modern male-dominated urban elites are so powerful that they often neglect the problems facing poor urban communities. However, it is not just patronage that reinforces the powers of mayors and public officials. Schneider, Teske and Mintrom (1995) refer to mayors, politicians and public officials as new 'public entrepreneurs', and Peters (1997, p. 55) argues that economic globalisation probably enhances the role of local governments, especially when they extend their interests with the public and private sectors. They have also developed regional partnerships linking to local communities (Orfield, 1997) and they have established themselves within new regional cultures of policy networking and enterprise development.

2 Strategy, Partnership and Contingency

Cultural theory provides a good contextual analysis of governance, but it does not detail the relationships between networks, public and private organisations and policy strategy. Cultural theory usefully identifies the social changes that contextualise change in government, but the relationship between organisational complexity and policy outcomes is not addressed. As the Ostroms (1997) argue, cultural theory currently provides a broad framework for analysis as opposed to a way of effectively dealing with specific policy problems. In contrast, a contingency model is employed in this chapter to establish a model for analysing the patterning of different economic development and urban regeneration partnerships. The model places emphasis on the importance of interorganisational relationships and the external and internal contingencies (Kouzmin and Jarman, 1989) that shape partnerships and influence strategy making. The chapter begins with a brief overview of extant urban theories and reviews major disagreements about the roles and characteristics of policy networks.

Urban theory and networks

The recent theoretical interest in networks, especially in Britain, followed the failure of macro, or nation state level, studies of government to address issues relating to the increasing complexity, remoteness and fragmented nature of public policy (Rhodes, 1997). Pluralist writers argued that many groups influenced policy. Critics argue this provided democracy with a political justification (Waste, 1986; Dunleavy and O'Leary, 1987; Judge, 1995). In his classic study of New Haven, Dahl (1961) argued that interest groups influenced different issues and 'polyarchy' allowed 'the many' to influence policies thereby avoiding elite domination. Political actors shared power and resolved conflicts to the mutual benefit of contending groups. However, Bachrach and Baratz (1962) defined the values and biases of political institutions and the 'non decisions' that influenced policies to the detriment of excluded interests. Even Dahl and Lindblom (1976) later recognised the problems encountered by social groups, expressing concern about social inequality and the deficiencies of the political system. Manley (1983) describes this, and Dahl's (1982) reassessment, as a fundamental revision of pluralism that shifted attention to new institutions and new concentrations of political power.

Lindblom (1983, p. 384) acknowledged that accepting 'the privileged position of business' represented a partial break from early pluralism and that contemporary politics was 'pluralist only on secondary issues.' However, such formulations represented the problems and contradictions confronting pluralism rather than any coherent theoretical reassessment. Neo-Marxists were providing more fundamental challenges to the old pluralist model as they focused on the role of communities in social reproduction and the ways in which institutions changed under modern capitalism (Castells, 1977; Harvey 1988). Regulation theorists regarded the pluralist revisionists as providing a wholly inadequate interpretation of modern capitalism and they took issue with many neo-Marxists over the nature of the capitalist crisis and the role of modern capitalist institutions. Regulation theorists argued that it was important to take account of the role of deeply structural capitalist crises that strained social and political institutions and undermined welfare liberalism and state collectivism (Jessop, 1994). However, there were different regulationist interpretations, with some explaining capitalist institutional transformation conditioned by the passing of Henry Ford-style mass production. The mass production and hierarchical political institutions that once regulated capitalism were giving way to economic flexibility and institutional fragmentation.

Amin (1994) argues that while post-Fordist theories analyse the systemic transformation of capitalism, regulation theorists remain deeply divided over the nature of capitalist restructuring (Amin, 1994, p. 3). Indeed, some regulation theorists have adopted more incremental approaches that do not rely on the post-Fordist periodisation, and this has produced 'a confrontation of diverse viewpoints, a heterogeneity of positions which draw on different concepts to say things about past, present, and future' (Amin, 1994, p. 5). According to some deterministic post-Fordist accounts, institutions reflect changes in the mode of capitalist regulation so that when production reorganises, so do public institutions. Mayer (1994) therefore refers to post-Fordist governance, although Jones (1997) questions such determinism by describing the irregularity of political relationships that link the central state to regional policy networks. Jones draws upon Jessop's strategic-relational state theory that reveals the selective nature of central state interventions and their impact at the local level. The 'hollowing out' (Jessop, 1994) of the state distributes central state functions downwards to subnational institutions and upwards to the global. Local governments and regional institutions are therefore more important in the development of the flexible economy as international bodies operate across boundaries in the global economy. This produces policy selectivity that favors certain regions, such as the South East in Britain, that contribute most to national economic growth and competitiveness. Jones therefore stresses the contingent nature of central policy and the consequent political competition between regional agencies and local governments.

Regimes

Like regulation theorists, regime theorists stress impermanence and variety in their rich insights into change in urban government. Regime theorists study power and the formation of coalitions and public–private partnerships, but Stoker (1995) shows considerable diversity in regime approaches. Following Judge (1995), Stoker likens regime theory to late pluralism because of the concentration on groups that work to 'accomplish public purposes' (Stoker, 1995, p. 55). Stoker also detects a neo-Marxist influence in regime theory with the stress on corporate political involvement. Stoker observes that regime theorists concentrate on governments that create the capacity to govern through social and political compromises with groups (Stoker, 1995, p. 66), but they assume that the private sector has a privileged position that often excludes community groups from the policy process. Regime theorists regard the lack of centralised political power in American cities as a defining feature of local government making it important for policy makers to cooperate with corporate and other interests to govern effectively (Stone, 1993).

Savitch and Thomas (1991) describe 'hyperplural' conditions where sustainable political coalitions fail to form because business and political elites lack sufficient power. The decline of central cities and the growth of suburban economies in the USA strengthens the 'hyperpluralist tendency' (Savitch and Thomas, 1991, p. 11) by dispersing power so that centralised authority in many cities dissipates. This spawns 'a centrifugal, undefinable, and ungovernable metropolis where decisions splay in many directions and add up to little' (Savitch and Thomas, 1991, p. 11), and where the political gravity of cities shifts away from formal public institutions to complex networks. By contrast, other regime studies find that competing groups can retain policy control especially when it is in the interests of governments and business leaders to establish viable regimes. However, while they emphasise the lack of political cohesion that weakens city governments so much, many regime studies provide valid comparative generalisations, and many regard urban governments as important despite extensive political fragmentation (Stoker, 1995, pp. 58–9).

Rational choice

The recognition of crisis and fragmentation in regulation and regime theories suggests that something appears to be fundamentally wrong with old-style hierarchical government as a mechanism for providing social and welfare services and that networking is one avenue for governments searching for new solutions. Against this, rational choice theory directs attention away from systemic crises to the actions of political actors in state bureaucracies inflated by high spending welfare liberals. Scharpf (1997), adopting a rational game theoretic approach, argues that traditional rational choice studies present political actors attempting to maximise benefits within rule bound institutions, and such studies develop a narrow institutional view concentrating on rules and systems of rules that organise behaviour (Scharpf, 1997, p. 40). Although the narrow approach in mainstream institutional rational choice neglects networks, Scharpf (1997) himself presents an institutional analysis that begins to redress that neglect by regarding networks as conducive arenas for actors making complex rational political choices.

Public choice and rational choice provide many insights into bureaucracy (Downs, 1967; Niskanen, 1971), but Dunleavy (1991, p. 4) argues that the 'conservative value-bias' in rational choice theory restricts its application. He argues that despite the bias, rational choice methodology is 'not intrinsically tied to right wing political values' or anti state theoretical assumptions (Dunleavy, 1991,

p. 5). Instead, rational choice can explain bureaucrats' motivations and can be applied to analyse the problems of modern governance in radical ways. Similarly, Dowding, Dunleavy, King, and Margetts (1993) suggest that rational choice can move urban theory forward. Following Dowding (1991), they claim that rational choice should go beyond pluralism, which theorised power as self-contained within communities, by viewing power exercised through interconnected networks. They argue that because local institutions distribute power in networked political systems rational choice theory can usefully explain the resulting complex power relationships. However, as Marsh (1998) argues, rational choice concentrates too much on the role of agents and their preferences and too little on the role of network structures as influences on policy outcomes. However, Scharpf's (1997) treatment of networking suggests that rational choice can usefully reveal the relationship between the preferences of political actors and the structures that develop within networks. This is especially the case in networks that structure to enable actors to achieve hierarchical coordination and political control. Such issues have been important in the formulation of theories specifically dealing with network interactions and patterns of network organisation. 'Network theories', however, tended to home in on the characteristics of networks themselves with inadequate emphasis on environmental contexts within which political actors work. This has led in network theory to a preoccupation with producing typologies that describe different network configurations and political relationships and with endless attempts to map networks and represent their components graphically.

Networks

Because network theories are structured around intergroup bargaining over resources, they naturally compliment game theoretic accounts of strategic positioning between actors (Scharpf, 1997). However, while network theories focus on the influence of power structures and dependencies between political actors, there are conflicting interpretations of how networks organise (Perrow, 1986) and disagreements about the usefulness of different typologies. There are also problems with neat diagramatic portrayals of networks that draw them as circles and lines, implying an exclusivity between their various parts. Figure 2.1 illustrates some possible ways of describing informal or atomistic networks and more complex decentred partnerships that cluster like starbursts (Quinn, Anderson, and

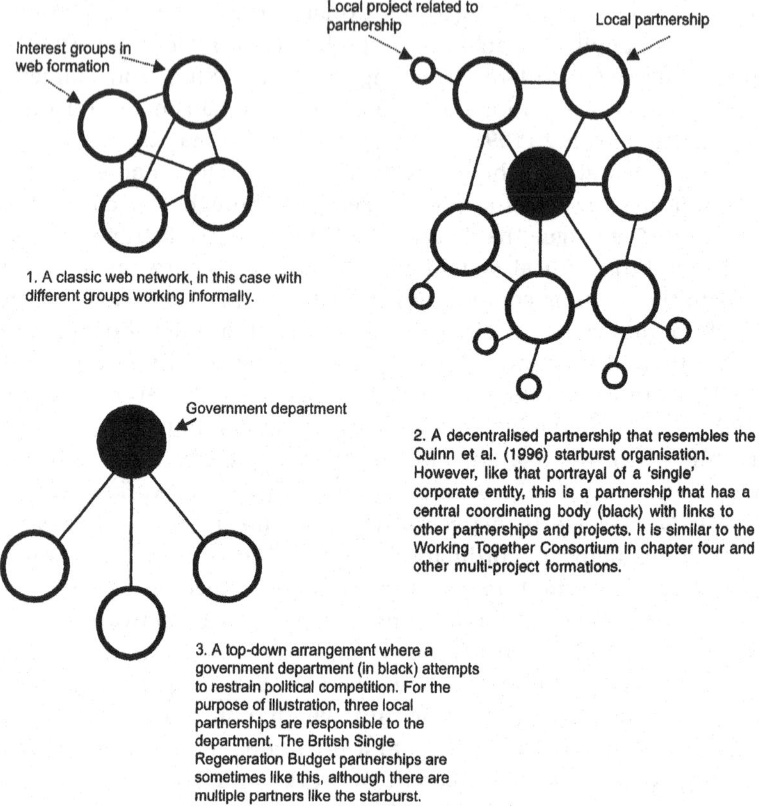

Figure 2.1 Classic Portrayals of Networks and Partnerships

Finkelstein, 1996). These are useful to aid discussion, but in reality networks and partnerships do not form into tidy categories.

Marsh and Rhodes (1992) adopt a network theory typology based on a structural explanation of network formation adapting a meso level perspective rather than macro level analysis. They acknowledge the contribution of North American literature that divides policy making into subsystems, and they cite Freeman (1955) who studied the growth of subgovernment and interconnected agencies and interest groups in the USA. Marsh and Rhodes especially draw inspiration from European approaches to interorganisational relations, and they therefore concentrate on relationships between institutions as opposed to personal relationships between politicians and public officials. However, they argue that in Britain, 'policy network' studies compliment European conceptions of interorganisational government (Marsh and Rhodes, 1992, p. 9) that are also relevant in the analysis of fragmented governance in the USA.

Marsh and Rhodes (1992) deal with both formal and informal networks, variety and ambiguity in relationships between groups, and attempts managerially to rationalise unstructured relationships. They refer to complexity and lack of pattern in networks, and Rhodes (1992) argues that policy makers encounter problems that 'factorize' or 'decompose' into semi-independent parts. Policies constantly change as they produce impacts in different parts of the government system and governments rationalise policy making through new management processes. This does not guarantee better outcomes because fragmentation generates disputes about overlapping jurisdictions, resources, and structures. Rhodes (1992) argues that rationalisation in government provides a distinctive focus because it decentralises big government and produces more plural policy. However, rationalisation strengthens policy networks by bringing groups together, and networks often close-up as 'the organized few' consult 'the disorganized many' (Rhodes, 1992, p. 60).

Dowding (1995), adopting a view that stresses the role of agency in networks, criticises Marsh and Rhodes for providing sets of labels that conceal the dynamic forces in networks that alter the balances of resources and power between organisations. Dowding regards the Marsh and Rhodes typology of networks as a descriptive tool that simply defines structural relationships at the expense of a proper appreciation of the role of agency in the form of actor preferences and actions. Dowding's assertion that the network typology is restrictive is valid, although his agency approach leads him to a narrow conception of actor roles. However, Dowding is right to argue that Marsh and Rhodes provide a formalistic typology evidenced by their reference to policy communities that have the 'continuity of a highly restrictive membership' and where groups have 'insulation from other networks and invariably from the general public' (Rhodes, 1992, p. 78). In this typology, 'issue networks' are the least integrated, comprising of a large number of participants with limited interdependence. Network structures tend to be 'atomistic' (Rhodes, 1992, p. 78) when groups work independently and haphazardly without policy focus, and since there 'is no one pattern of relationships for all policy areas' (Rhodes, 1992, p. 81). Policy communities are the most integrated and stable policy networks, but the assumption of stability and continuity within them limits the consideration of the disorderly processes of organisational–structural development, competition, and conflict within networks. Marsh (1998) in his later work better deals with discontinuities in policy networks and he sets an agenda for researching the ambiguities of network organisation that are the result of change in the external environment. The need for a better appreciation of the context of networks is important given the fuzzy boundaries between networks and governments and the resulting problems confronted by groups seeking effectively to make the best strategic choices about which networks they should join (Dudley and Richardson, 1996). Such renewed theoretical interest in the strategic choices of actors working in complex environments suggests the importance of a better appreciation of the processes of change in policy networks that result from a combination of internal network politics and the restructuring of external institutions and state structures.

This need to account for dynamism and change in networks is underscored in Dutch work on networks by Kickert and Koppenjan who regard policy networks more comprehensively as part of a new governance where hierarchical central control does not work because networks have 'no top' (Kickert and Koppenjan, 1997, p. 39). This perception tilts analysis to pluralism, since centralised management does not work because actors in networks have to participate with others to achieve their goals. Actors have to work together in 'cogovernance' to produce mutual advantages and positive gains (Kickert and Koppenjan, 1997, p. 40). In this scheme, networks are self-governing with groups entering into numerous dependency and resource relationships.

Knoke, Pappi, Broadbent and Tsujinaka (1996) also present networks in a wider context, adopting an interorganisational approach based on a resource exchange model. They describe a modern organisational state in advanced industrial societies where 'highly differentiated social, political, and economic structures' exist. Knoke *et al.* identify organisational ambiguity in political power relationships and within 'numerous relatively autonomous policy domains' (Knoke *et al.*, 1996, p. 9). In the 'organizational state', policy domains mobilise actors with common policy interests, but groups do not necessarily have identical policy preferences as they divide their attentions across domains. Therefore, 'domain boundaries are more or less fuzzy and porous, allowing various participants, problems, and policy proposals to enter and leave in disorderly fashion' (Knoke *et al.*, 1996, p. 10). Sabatier (1993) also views policy subsystems that change as actors become active and form advocacy coalitions. Advocacy coalitions contain 'actors from a variety of governmental and private organisations at different levels of government who share a set of policy beliefs and seek to realise them by influencing the behavior of multiple governmental institutions over time' (Sabatier and Jenkins-Smith, 1993, p. 212). Different coalitions may gain control of different parts of the government system or government departments (Sabatier, 1993, p. 28).

A contingency model

The structural focus of early network theory, with the emphasis on resource dependencies and stable relationships, therefore overstated continuity in the policy process. In addition, discussion about the role of strategy and how networks position themselves in the wider external environment was limited by the preoccupation with internal dependencies and network regulation issues. There was therefore a need, recognised increasingly by network theorists (Hay, 1998; Daugbjerg and Marsh, 1998), for an approach that better put networks into context by accounting for both their internal dynamics and the external environmental influences that shape structures and strategies. Arguably, there is also a need for an organisational frame to advance the discussion of complexity and ambiguity in the organisation of partnerships. The contingency model presented here is appropriate to meeting these needs because it integrates the analysis of strategy and structure, and it removes hard analytical lines between types of network and organisation.

The contingency model moves away from a rigid typology not simply because rigid definitions of different network configurations do not work, but because such typologies restrict our understanding of networks as forms of organisation. The contingency model establishes an organisational perspective of networks that, in a sense, over-rides popularly used and formalistic typologies. 'Unpacking' the terms 'network' and 'partnership' thus reveals the problem of over reliance on formalistic categorisations and it surfaces the need for a reassessment of networks accounting for their tendency to continually restructure and produce hybrid forms. Therefore, while categorisations may reflect popular perceptions, in theoretical terms 'networks' and 'partnerships' are actually to do with the characteristics of ambiguity and complexity that are typical of multi-interest forms of organisation. However, in the analysis that follows the focus is on subdomain and subsectoral forms of interorganisational collaboration. Therefore, the focus is less on the broader policy domain level 'policy networks' that feature so often in British network theory literature. It is more about subsectoral networks and partnerships and their indistinct boundaries. The analysis views partnerships that are often set within the wider policy network universe but which operate very much at the regional and local levels. Sometimes these partnerships grow and behave like larger 'policy networks' when they lobby governments and articulate interest group demands and expand their domain coverage (like some big regional partnerships), but usually they strive for more specific goals by being nested within the wider policy domain (Knoke *et al.*, 1996) Partnerships in economic development generally, although not exclusively, thus tend to specify their objectives to achieve particular stated aims but they broaden out over time and possibly become more aware of the need for wider strategic vision. Therefore, the divisions between formal partnerships and informal networking are not hard and fast since small networks expand, partnerships develop and policy horizons often expand from the local to the regional.

To ilustrate this, it is useful to consider informal local networks that 'organise' on the basis of unwritten understandings and loosely defined objectives. Groups within a network commonly declare a 'partnership' when the network develops more formal procedures, structures and competencies. Partnerships in economic development therefore are usually formed for specific purposes with stakeholders often adhering to a 'shared vision'. Such 'policy partnerships' typically are based on formal joint working agreements between local governments, public agencies and private groups for social and economic purposes or to promote competitiveness. However, some 'partnerships' are simply commercial agreements between organisations. They are not strictly 'informal networks', but are based on agreements between parties usually made under statute law and with parties agreeing to bear defined risks. Other partnerships between governments and the private sector are essentially industrial partnerships (Pierre, 1998) that support, for example, high technology applications or which concentrate on the diffusion of technologies. Industrial partnering is sometimes accommodated within wider regional policy partnerships, as in the Pittsburgh region, or at the local level through organisations such as the Pittsburgh Regional Alliance (see chapter four).

In policy partnerships, as opposed to those that are specifically commercial or intended to facilitate a financial deal, an initially loose network often provides the basis for the later development of formal and complex arrangements with hierarchical organisation and common services for members. While the analysis below concerns policy partnerships, structural ambiguity alerts us to the possibility that policy partnerships often embody commercial 'partnerships' that nest within (Perrow, 1986). Such ambiguity in partnerships requires a model that accounts for a range of political relationships and organisational arrangements. The model is inteded to account for situations where partnerships nest within other partnerships, where organisations adapt to changing external conditions and where diverse political relationships influence policies and the strategic commitments of the stakeholders. The diversity of patterns of organisation suggests that such arrangements are heavily contingent on the particular circumstances that condition partnerships and that policy strategies also need to be understood in this way.

The contingency model: propositions

The model derives from research (Jacobs, 1993, 1996) based upon core assumptions that originally identified network characteristics in terms of bureaupolitical relationships with public bureaus as the focal organisations for analysis. The model below is applied specifically to partnerships that are quite narrowly defined in policy terms, but which produce a wide range of public–private sector interactions. To that extent, the model portrays partnerships as organisational phenomena, and it is not intended to be a general theory of governance. The core assumptions underlying the model are as follows.

In the wider policy domain, governments and groups do not exist in equilibrium. Competitive and unstable relationships between private groups, governments, public agencies and partnerships result in change and uncertainty. Public–private partnerships especially present a range of complex organisational problems that often magnify those within traditional public institutions. For example, partnerships are managerially problematic, organisationally diverse and susceptible to change.

Politics pervades relationships between governments, public agencies, private and nonprofit organisations. Political competition influences the interactions between politicians, public officials and corporate executives, so policy makers strive for order and try to restrain political competition to avoid conflicts by incorporating various groups into the policy process.

The relationship between policy strategy and the organisation of partnerships is complex. The model emphasises the importance of strategy and shows that there is no straightforward relationship between policy strategies and organisational structure in public–private partnerships. Policy strategies influence partnership structures, but broader external macro level contingencies force political changes and produce unintended outcomes. External contingencies influence organisations in a variety of ways according to specific national and local conditions. National policies therefore provide one context for partnership working and for the patterning of partnership structures. Regional and local partnership strategies, while important in terms of market positioning and policy, frequently reflect nationally determined templates for partnership structural development, but regional partnerships are also influenced by internal factors such as political competition and by external contingencies. Regional and city strategies produce very changeable and subtle influences, being to greater or lesser degrees salient depending on the particular conditions, risks and turbulence prevailing in the policy system.

Networking encourages diversity in regional and local organisation, but the distinctions between organisations are blurred. Organisational diversity and ambiguity arise as local governments and partnerships develop new and innovative structures to address problems internally and in their external environments. It is therefore necessary to account for this diversity in the model.

The following expands upon each of these aspects of the model.

Disequilibrium

Neoinstitutional writers, March and Olsen (1989), describe internal and environmental factors and the stable and routine processes that mold institutions. Their contention is that 'if the environment changes rapidly, so will the responses of stable institutions' (March and Olsen, 1989, p. 58). For them, the 'institutionalization of action' (March and Olsen, 1989, p. 21) consolidates rule-bound order in organisations, although they concede that even rule-bound institutions change and that policy failures produce structural changes, organisational learning, and reassessments of past actions. Kouzmin and Jarman (1989) question the viewing of organisational change in terms of stability, continuity, and order. If change implies a move away from 'positivistic certainty' (Kouzmin and Jarman, 1989, p. 398) that means that decision makers do not always manage to control events. Policy makers like to think that they are in control, but they do not always succeed. Jarman (1994) argues that institutional rules may provide order, but policy makers break the rules especially when they confront external factors that undermine their best laid strategic plans. Jarman thus doubts conventional accounts of change that concentrate upon organisational effectiveness and equilibrium leading to growth. Following Thompson (1967), Kouzmin and Jarman utilise contingency theory to argue that the strongly held myth of equilibrium leads policy makers to rely in practice upon routine, planning and control because they mistakenly assume that organisational growth is natural. Policy makers believe that control and order underpin successful change, but the policies based upon this myth fail when policies are disrupted. Therefore, Kouzmin and Jarman argue for a fuller theoretical and practical recognition of cognitive failure and disequilibrium when accounting for policy change and organisational innovation. This is particularly important where policy makers face unique situations that require nonstandard or crisis responses (Jarman, 1994). For example, urban riots provide an extreme example (Quarantelli, 1993; Baldassare, 1994) because social disorders have unique characteristics that require fast policy responses. Politicians and public officials attempt to control the political process and restrain the conflicts between political actors, but uncertainty and unpredictability encourage solutions that are politically possible and practically workable. Mintzberg (1996a) regards change as providing managers in organisations with the opportunity to learn from experiences and adopt policies that relate to external conditions. Mintzberg, challenging conventional wisdom, suggests that the most effective managers are those who can deal with change by 'crafting' strategy through 'thought and action, control and learning, stability and change' (Mintzberg, 1996a, p. 107). In this way, strategy is one of the control mechanisms of organisations that reduces risks and increases stability.

Political competition

Change, or the threat of change, politicises strategy because it upsets relationships between groups in policy networks. Mintzberg (1996b) argues that there is a tension between change in the external environment and the expectation of ordered strategy. Rosenthal, Hart, and Kouzmin (1991), adopting a contingency approach, suggest that public officials achieve different degrees of policy control and that plans and strategies often collapse under the pressure. Some friction between organisations is beneficial when it produces new ideas and improved administrative structures, but dysfunctional conflict leads to the breakdown of good relationships between groups and disordered decision making. Multiformity (Godfroij, 1995) in networks assumes the involvement of groups with differing interests where actors differ about their commitments to networks depending upon their representational and resource bases. Therefore, even when there is a mutual agreement between groups, partnerships are not necessarily stable because actors assert their own interests and compete for public resources. Partnerships are particularly prone to conflict over the sharing of operational, financial and other risks and responsibilities (Bunyan, 1994). However, Scharpf (1997) shows different relationships between groups that range between solidarity where groups cooperate and individualism that makes collaboration difficult. For Scharpf (1997, p. 86), relationships also can be altruistic or hostile, and circumstances condition the types of relationships between groups.

Rosenthal *et al.* (1991) describe the tendency for political actors to work in disorderly and unpredictable ways. They describe relations between public agencies that range between a restrained competition, where conflict is unimportant or controlled by policy makers, to 'bureaupolitism' where conflict is dysfunctional and out of control. Rosenthal *et al.* (1991) apply their approach to the broad policy domain, but it is also relevant within particular networks and within individual partnerships. Rosenthal *et al.* (1991) signify 'degrees' of political restraint and political conflict and the factors that influence political conflict. Indicators measure gradations of bureaupolitical competition which means there are no fixed demarcations between stable and conflictual situations. For example, if there are many actors, a positioning of interests, and intergroup power struggles, then compromise is difficult to achieve, and conflict may be dysfunctional. If conflict between groups is rife, it is difficult to meet policy objectives through an effective strategy. The combination of variables differs according to circumstances producing a rich variety of possible outcomes. Policy makers therefore have an interest in ordering the policy process and achieving group co-operation. Bureaucrats work with politicians and others to produce workable compromises to implement policy strategies. However, bureaupolitism can indicate dysfunctional conflict where agencies have different objectives and openly express their differences. In contrast, bureaucratic politics, in its generic form, assumes many competing actors attempting to influence policy, but different agencies reach workable compromises in spite of political competition between them. For Rosenthal *et al.* (1991), restrained competition produces comparatively 'simple' relationships based on compromise and consent. Restrained competition resembles the Marsh and Rhodes (1992) policy community, but unlike the policy community it implies an inherent tension between order and potential political competition within networks. Restrained competition is at the other end of the continuum to conflict because restraint is conducive to order as policy makers try to avoid disagreements and conflicts. The policy community has 'consistency in values, membership, and outcomes which persist' (Rhodes, 1997, p. 43), but restrained competition implies tension and changing intergroup relationships even in situations that resemble stable policy communities.

The Rosenthal *et al.* (1991) model is useful because it describes graduations of political restraint and conflict. However, there are problems to the extent that their model it is specifically focused on governmental organisations and therefore de-emphasises the roles of private groups. Second, the model assumes that larger numbers of actors in a policy domain or network produce bureaucratic politics and smaller numbers are likely to lead to restrained competition. This does not necessarily hold in economic development partnerships where large numbers of actors may be politically restrained. Third, restrained competition, bureaupolitics and conflict exist in their model in relative isolation from the organisational arrangements that 'support' such different patterns of political interaction. Rosenthal *et al.* describe the degrees of bureaucratic politics and conflict, but they neglect the organisational arrangements that 'support' different degrees of political competition. The bureaupolitical model therefore needs an organisational dimension because as it stands the indicators of political competition only suggest the possibility of different network structures and organisation. They provide only a vague indication of possible organisational forms that accompany different degrees of political competition.

However, by profiling the organisational dimension, and by employing a modified version of the Rosenthal *et al.* (1991) political competition–conflict continuum (Figure 2.3), we can enrich the analysis of networks. The organisational dimension enables us to view politics as played-out between public and private organisations in various networks and partnerships where there develop different structures that shape according to the outcomes of political bargaining and compromise. Viewing political competition and organisation together helps to break down hard typological distinctions between formal and informal networks, partnerships and hierarchical organisations. An interorganisational–political focus reveals the wide variations in structures and political competition within networks and it is consistent with approaches that have drawn on organisational theory as well as political science (Klijn, 1997).

Strategy and organisation

To appreciate the organisational aspects of partnerships, it is useful to see what influences their structures. The factors that influence the structures of networks and partnerships are found within the external environment (see chapter three) and within the networks themselves. Heuvelhof and Bruijn (1995) suggest that some external events assist urban policy makers while others, like riots and factory closures, hamper their efforts and threaten old policy commitments and institutions. The dynamic external organisational environment forces policy makers to make strategic choices that disrupt prevailing norms and lead to new structural arrangements. Goetz and Clarke (1993) acknowledge that in local government new opportunities arise under such fast changing conditions. Clarke (1993) refers to Tarrow's (1991) 'international opportunity structure' when new opportunities arise for private organisations and governments to work together and develop new policies. External change disrupts long-established structures, but policy makers initiate new strategies and organisational solutions. Change can be dysfunctional and disruptive, but it can also bring strategy and reorganisation.

Quinn (1996, p. 3) regards strategy as a factor that influences organisations in complicated ways. He argues that strategy is 'the pattern or plan that integrates an organisation's major goals, policies and action sequences into a cohesive whole. A well formulated strategy helps to marshal and allocate an organisation's resources into a unique and viable posture based on its relative internal competencies and shortcomings, anticipated changes in the environment and contingent moves by intelligent opponents'. Policies are 'rules or guidelines that express the limits within which action should occur. These rules often take the form of contingent decisions for resolving conflicts among specific objectives.' (Quinn, 1996, p. 4). Strategic policies are major policies that influence the direction of an organisation, its market position, and its ultimate existence. 'Strategy making' however involves a combination of both 'position' and 'perspective' (Mintzberg, 1995a, pp. 17–19). Position is about finding an appropriate orientation within the external environment, and policy makers have distinctive views of the world that influence strategic choices and policy commitments. They interpret their environments from their own particular perspectives and devise strategies to see them through risky situations. Partnerships have many actors, so shared strategic visions usually represent the agreement of stakeholders to work around common goals even though they may retain their own perceptions of change (Connell, J. P. and Kubisch, 1997). This means that strategy is a compromise between different interests that agree to focus their common goals within the partnership and move beyond incremental policy responses.

While strategy is important in providing direction, researchers have tried to describe the nature of its impact on structure. Mintzberg and Quinn (1996, p. 320) assert that organisational structure 'no more follows strategy than the left foot follows the right in walking'. There is no straight line relationship because strategy and structure are interdependent. They maintain that the choice of any new strategy is 'likewise influenced by the realities and potentials of the existing structure' (Mintzberg and Quinn, 1966, p. 320). Galbraith (1995) argues that strategy influences structure, but other factors can influence structure even more. Information and decision making processes within organisations often force change, especially when there are technological innovations. Human resources and systems of rewards change, and external relationships influence internal processes (Galbraith, 1995, pp. 134–5).

In regional policy, there is the complication that overarching regional partnerships promote strategies that often do not match with the strategic perspectives and commitments of local partnerships. In Pittsburgh, Birmingham and Rotterdam, regional fragmentation often produces competing strategies which complicates the process of partnership development, hinders the attainment of shared visions and, as in the USA, forces some partnerships to adopt structures more appropriate to their local needs. In urban regeneration, central government departments and city councils adopt new strategies that sometimes only marginally affect established basic partnership structures even though policy direction may change significantly. The analysis in subsequent chapters therefore treats regional and city strategies as contextual influences for organised partnerships that work within such wider policy frameworks. The case studies show the intricate interrelationships between broad policy strategies and local partnership programmes and the subtlety of interorganisational relationships that arise from a variety of changing contingent variables. The cases also show that 'partnership' is really an aspect of organisational diversity and that 'partnership' itself is not a static type.

The study of 'visioned' strategies and partnerships in urban regions reinforces the view that there is no single organisational structure that consistently goes with any particular strategy, and that there are complex interactions between strategy and structure. City governments in Rotterdam, Pittsburgh and Birmingham support partnerships that have various organisational structures that are both appropriate to local conditions and which can conform to centralising blueprints. New structures often add to diversify in administration and service provision (Halachmi, 1995), sometimes by devolving activities to the local level, sometimes centralising programs. Mintzberg (1994, 1996b) provides a configuration model of this organisational diversity that builds from contingency theory (see also Miller and Friesen, 1984), and he views organisational diversity as one outcome of the choices made by policy makers in complex environments. He argues that internal structure, tradition, technology and culture differ between organisations, and these play a role in shaping structures and policies. Older organisations, and those that operate in stable environments, tend to adopt formal control and hierarchical structures of organisation, but in environments that are more complex and unstable, organisations compete with one another and rely more on informal relationships. Innovative structures develop as competition forces policy makers to form alliances and decentralise operations to maximise their competitive advantage and capture new markets.

Recent studies of competitive private corporations indicate that policy makers develop innovative and flexible strategies, processes, and structures effectively to respond to global competition (Porter, 1985; D'Aveni, 1994; Hamel and Prahalad, 1994; Gouillart and Kelly, 1995). The participation of city governments in networks also facilitates internal reorganisation and process change (Pierre, 1998). Alterations to management processes and core structures achieve internal consistency within corporations (Davidow and Malone, 1992; Nohria and Berkley, 1994; Bowman and Kogut, 1995; Galbraith, 1995; Yoshino and Rangan, 1995) and in public organisations. The organisational models developed by Mintzberg (1996b), Clegg (1990), and Galbraith (1995) all indicate how networks continually reconfigure and how networking with external organisations contributes to this adaptive process as policy makers strategically reorientate and re-position their organisations to make an impact in the competitive environment (Genus, 1995; Scott, 1995).

Partnerships as developing organisations

Mintzberg (1996b) defines the basic parts of an organisation that develop over time. These essential building blocks are found in corporate and public sector organisations, although he does not develop their use in public–private sector networks. For Mintzberg, an extant organisation consists of an *operating core* that delivers the service or product. A *strategic apex* commonly consists of a group of managers who oversee the running of the organisation. A *middle line* of management usually develops as the organisation grows. The middle line links the strategic apex to the operating core through a line of management authority. The organisation often develops a *technostructure* that exists outside the line management hierarchy and consists of the analysts who plan and control the work of others within the organisation. There develops a *support staff* providing specialist services such as legal advice and mail room operations. Finally, an *ideology* develops that embodies the beliefs and traditions of the organisation.

By borrowing these terms for multiorganisational networks, it is possible to identify similar organisational functions. Yoshino and Rangan (1995) refer to the similar organisational imperatives of single organisations and multi-interest networks. Many of the partnerships in economic development studied in the following chapters consist of interconnected and specialised parts that are very like those in the Mintzberg organisation. The administrative subdivisions of partnerships sometimes operate like departments in a large organisation. For example, a core of managers or a board might make strategic policy and oversee the partnership's operational activities. In Mintzberg's single organisation, divisions link internally through the management structure with the technostructure and support staff. In interorganisational networks with many actors, there is a need to develop core competencies that resemble those in single organisations. Organisational functions and competencies resembling those that develop with the 'parts' of Mintzberg's single organisation may therefore also develop in inetrorganisational networks and partnerships (Yoshino and Rangan, 1995). The level and type of activity in the partnership produce higher or lower degrees of organisation within the partnership depending on the circumstances. In a sense, partnerships try to emulate single organisations that have many 'enviable' core competencies. As the internal administrative functions of a partnership expand, new groups join and more sophisticated methods of coordination between them need to be developed. Partners create a more effective organisation through specialised administrative arrangements, and as the partnership develops the distinction between 'networking' and 'organisation' blurs. Initially informal networks therefore develop stronger organisational characteristics as they grow and become successful. Alternatively, some hierarchical organisations decentralise to benefit from the flexibilities of looser networks, but retaining strong internal communications and control. Public and private sector organisations therefore develop varied patterns of organisation and strategy that are appropriate to handling the complex problems that they confront. Different organisational structures and policy strategies result because policy makers attempt to achieve the most appropriate fits between organisational structure and the external environment. Ideally, there should be internal consistency.

Figure 2.2 applies a Mintzberg-influenced organisational approach to identifying the characteristics and 'parts' of partnerships using a British urban regeneration initiative as an example of one kind of complex configuration that can arise in a multiagency partnership. Challenge Fund initiatives are partnerships funded by central government, but organised at the local level. In the hypothetical example in Figure 2.2, which portrays a possible partnership arrangement, there is a multipartner strategic apex that subdivides to a managerial apex. The managerial apex, in turn, oversees line managers concerned with the operational activities of the partnership. There are local projects run in communities that deliver services designed to effect various improvements and empower local people. These projects link, starburst-style, to other projects and partnerships outside the Challenge Fund partnership. Program evaluation is performed through a technostructure where expert is advice provided by consultants. If the partnership can be kept flat and flexible, then there is less likelihood of a top–down management style, and local projects will thrive within the overall structure. However, it is very difficult to present such relationships in a diagram because, as in this case, the strategic apex connects extensively with the operational base

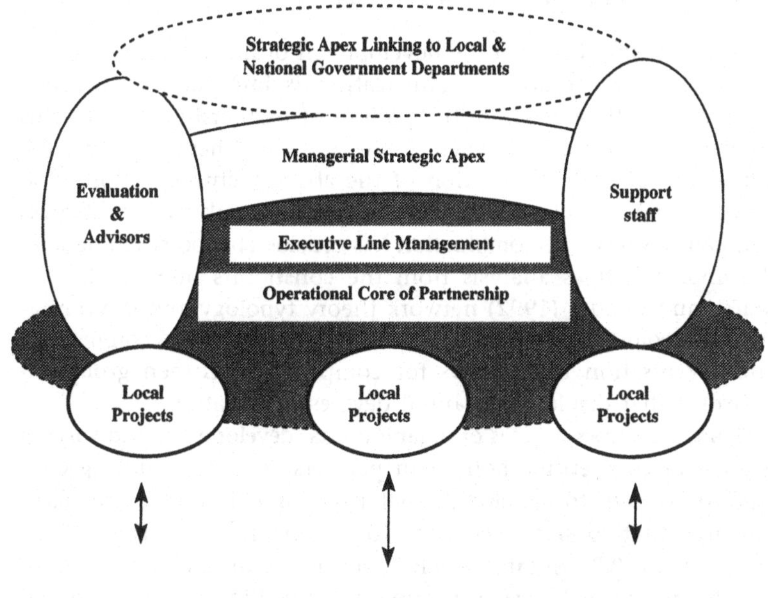

Figure 2.2 Hypothetical Urban Regeneration Partnership

and local projects. Also, the example suggests that in public sector situations there may be organisational 'parts' that represent the political necessity of ensuring democratic accountability to communities and local government authorities.

Figure 2.2 provides a more detailed impression of this kind of compact partnership than any of the classic views in Figure 2.1. Nevertheless, the classic portrayals can still provide useful general shorthand descriptions such as the starburst. What is most important is the recognition that the different parts of organised partnerships develop in different ways and to different degrees, producing structural diversity and local variation (Morgan, 1986). Quinn *et al.* (1996) show the infinite number of ways in which organisational parts develop within the 'network organisation' and how the distinction between network and formal organisation can be very blurred. They find that 'bureaucracy' develops when networks structure along organisational lines and that organisations in 'hypercompetitive environments will be hierarchical' (Quinn *et al.*, 1996, p. 352).

Joining-up organisation and politics

The relationship between the development of hierarchies of various kinds and organisational sophistication is important as the two interact, and they generate the politics of partnership. It is in this connection that the contingent model presented here combines the pervasive political competition of the above mentioned Rosenthal *et al.* (1991) model with the organisational analysis of networks and partnerships. The combination has considerable potential because the approach frees analysis from the constraints imposed by the Marsh and Rhodes (1992) network theory typology and the limitations of neoinstitutionalism. The combination of political competition and organisation thus allows for competition between groups in networks that display substantial degrees of variation.

Figure 2.3 shows degrees of organisational development and varying degrees of competitive politics in partnerships. The resulting configurations help to explain the extensive and rich variety of partnerships that exist in economic development and other policy areas. Figure 2.3 contains a left hand axis that shows a range of possible positions between informal relationships and more formal organisation and partnership. The axis represents different degrees of organisational development across the range of strategic apex, operating core, middle line, staff and technostructure provided in Mintzberg's framework. The more these organisational parts are developed within an emerging partnership, the more 'organisationally developed' will be that forming partnership. So, if a partnership establishes a strong administrative capability with a developed strategic apex linking partners together, then the partnership will resemble a line-managed organisation, and it will be more highly developed. If members of a network get together informally, then there may be little organisation to speak of, although they may later declare a partnership and begin to establish closer links and sophisticated administrative mechanisms. The location of a partnership along the continum showing the degree of organisation could be arrived at by allocating values according to the development of the parts of organisation (see also Chapter 10). The more developed the organisation, the higher the allocated score on the axis. In a similar vein, Figure 2.3 also shows a horizontal axis inspired by Rosenthal *et al.* (1991). This axis tracks between restrained competition and conflictual, usually dysfunctional, relations, and simply measures degrees of restraint and conflict without reference to the numbers of actors involved in policy (unlike Rosenthal *et al.* 1991). The two axes in Figure 2.3 therefore together indicate the potential variability of partnerships since partnerships can change along both axes. This provides for an infinite variety of combinations of political competition and patterns of organisation in partnerships (see chapter ten for a more detailed statement of this approach in the light of the case studies).

Figure 2.3 shows a situation of a hypothetical informal partnership arrangement that attracts more members and develops more complex tasks over time. The line A–B shows the developmental trajectory of the collaborative situation. The partnership initially is informally organised with few administrative functions (point A). At point B, the partner members declare the formation of a partnership, and they dedicate funding for administrative tasks carried out by the partnership. The partners manage to restrain competition and reduce internal conflict by working in line with a shared strategic vision. Point C represents another pattern with low organisation and high internal conflict between members. This could be the kind of situation that exists when conflicting neighbourhood groups are unable to agree on common goals. The partners may get stuck at point C and eventually break up.

Quinn *et al.* (1996) refer to different organised configurations that help to refine this approach in line with Figure 2.3. There are many different ways of describing different configurations and it is not absolutely necessary for network members to declare a 'partnership' just because they enjoy a high degree of organisation. Moreover, the Quinn *et al.* (1996) approach usefully shows that in competitive situations organisations restructure using managerial control in different contexts. For Quinn *et al.* (1996, pp. 350–62) there are many organisational–network possibilities. One is *flatness*, where informality exists in the absence of developed administrative and management function. Another is *inversion* that occurs where an organisation concentrates power and resources in its decentralised nodes, thus deconcentrating power through 'networks.' A *spider's web* forms if many sponsor organisations agree to collaborate with little hierarchy and widely shared risks. *Clustering* is like the spider's web, but where team members develop more specialist administrative and policy activities and cluster managerial competencies as a result in different nodes.

Starbursting creates a situation where members agree jointly to pool resources (Quinn *et al.*, 1996). Figure 2.1 showed a classic and over-

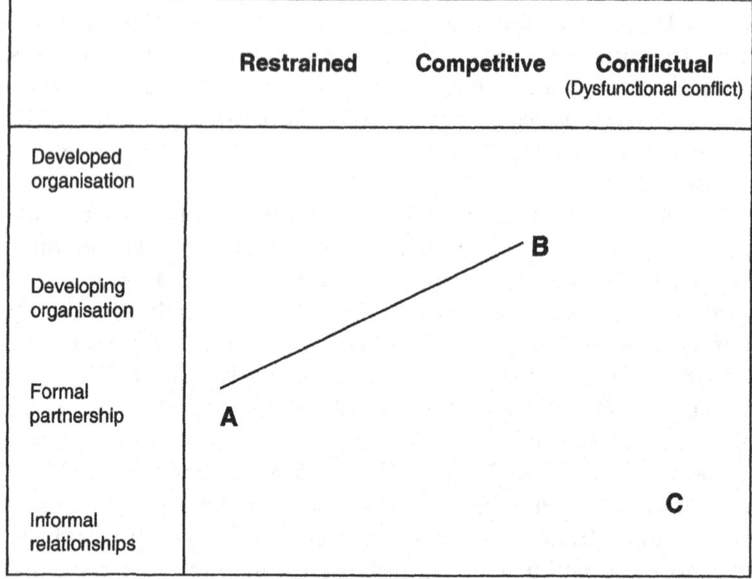

Figure 2.3 General Form of the Relationship between Political Competition and the Degree of Organisation in Partnerships.

stylised portrayal of such patterning. However, even such a stylisation indicates that partnerships often create and 'own' a specialist body to serve and administer a partnership. Here a strategic apex is granted some autonomy by the sponsor members of the partnership so that it can get on with specialist tasks. In economic development, the starburst and atomistic or inverted patterns (Figure 2.1) are common, although there is restraint when public organisations exert centralising control. The case studies in the following chapters show organisations that coordinate partnerships in starburst fashion, like the Urban Redevelopment Authority in Pittsburgh, and which themselves have complex structures and nodal relationships linking partners. Some public sector organisations therefore have similar functions to the core bodies within private sector starburst networks. In spite of the nature of 'network organisation' and the influence of new technologies, managers seek political restraint (Rosenthal *et al.*, 1991) to deploy staff in new ways to maximise resources, managerial flexibility, and the effectiveness of their market interventions (Quinn *et al.*, p. 352). Even the 'spider's web' has to 'connect' once a 'project forces it to materialize'. The model thus suggests a large variety of partnership structural possibilities, but also allows for hierarchical control of even the most diverse partnerships.

Richly textured partnerships

Figure 2.3 shows a typology that is flexible enough to accommodate wide variety in the comparisons of economic development and urban regeneration partnerships. The relationship between structure and political competition provides a useful picture of varied organisational forms. Combined with an appreciation of the political cultures within which diverse networks form (chapter one), the analysis suggests some very richly textured organisational and political arrangements. Partnerships thus vary according to the political contexts within which they form (chapter one), the ways in which they respond to their external environments (chapter three), their institutional links, the norms and values of public and private organisations, and the strategic policy aims that policy makers prescribe (Halachmi and Bouckaert, 1995).

The pattern in Figure 2.2 is relevant in economic development and urban regeneration, but it shows only one possible structure with a developed organisation. Also, partnerships usually deploy smaller resources to their internal organisations than corporations. They tend to develop less complex internal divisional arrangements. The Mintzberg configuration model provides a useful view of organisations for such purposes, but it under-emphasises political contingencies and over estimates the ability of strategy makers to achieve internal consistency. The Rosenthal *et al.* (1991) model places political contingencies centre stage. Hence, Figure 2.3 suggests that partnerships are less likely to find internal political consistency than the Mintzberg model would imply. Economic development partnerships do not always find consistency between management control mechanisms and structure. Often control has to be centralised by governments to reduce political conflict and overcome fragmentation. Politics cuts accross all types of partnerships and suggests that frequently partners make do with what is politically feasible even if it goes against strategic good intentions. Figure 2.3 is thus, in a sense, more politically contingent than the Mintzberg model and less likely to lead to best solutions.

3
Strategy and the Competitive Environment

The previous chapter described the relationship between strategy and the organisation of partnerships. The contingency model indicated that organisations vary according to their internal politics and conditions in the external environment. In regional economic development, external factors and network dynamics therefore influence partnership structures and city government strategies to produce complex and varied organisational configurations. This chapter concentrates on some major risky and unpredictable external economic environmental factors that influence public and corporate strategies and encourage public and private organisations to collaborate. The case studies in subsequent chapters cover the real workings of partnerships. This chapter headlines the economic contingencies that influence policy makers who formulate strategies. Together with social and political factors, these determine the issues that will be addressed by strategy makers. The chapter views economic variables that generally make life difficult for strategy makers and lead them constantly to revise their commitments, expectations, and ways of working. Consequently, strategies often count for less than garbage-can politics (Kouzmin and Jarman, 1989). Then, strategy is over-ridden by the short term expediency of politicians and public policy makers.

Formulating economic development strategies

Quinn's (1996, p. 3) claim that strategy is the pattern or plan that integrates 'goals, policies, and action sequences into a cohesive whole' implies a substantial task for private sector organisations and city governments trying to deal with change and uncertainty. For a strategy to succeed in business, a corporation needs to allocate resources so that it can adopt a 'unique and viable posture' (Quinn, 1996, p. 3) and develop core competencies in a competitive market. Public officials in Pittsburgh, Birmingham, and Rotterdam also seek to develop organisational and economic regional competencies that contribute to the economic well-being of cities and regions. Public officials and corporate executives therefore have a common interest in keeping ahead of the competition. Corporations want to make profits and expand into new markets, and city governments want local communities to thrive through growth and high employment. City governments therefore support corporate restructuring by backing new technologies and encouraging innovation and investment (Wadley, 1986; Pierre, 1998). They seek to develop the conditions under which transnational corporations can boost company competitiveness by using the human and physical resources of regions and localities. As Smidt (1992) argues, regions gain when locally based corporations achieve capital growth, enhance international standing and create wealth. However, the changing external environment often presents public policy makers with a series of moving targets making it difficult to settle strategy. This suggests the need for policies and strategies that are adaptive and responsive under changing conditions, although excessive attention by policy makers to painstakingly refining strategies often results in policy inconsistences, misinterpretations of the pace and nature of change, and a proliferation of dysfunctional shared visions and statements of intent. This is not surprising given the range of major variables that influence regional economic development strategies as indicated in Figure 3.1. These variables can change quickly, making even the most well conceived strategies redundant. Figure 3.2 shows how things can go wrong when unexpected events push policy off course and produce unintended outcomes. Figure 3.2 depicts a given 'settled' strategy that is disrupted, as may be the case if a region suffers a sudden collapse of an industry or major company that had been instrumental in the original strategy. For example, the possibility in 1999 that BMW could close the Longbridge automobile plant in Birmingham threatened to undermine a regional strategy that envisaged Birmingham as a hub for auto manufacture.

It is useful to focus on some of the economic variables that influence the policy strategies of city governments and their private sector partners because economic factors have direct short and long term implications for strategy makers. Chapter one dealt with the

Figure 3.1 External and Internal Factors that Influence Public–Private Sector Regional Economic Development Partnership Strategy Formulation

External Factors that Influence Partnership Strategy	Internal Factors that Influence Partnership Strategy
• external dependencies	• internal dependencies and organisation
• relationships with external interests	• relationships between partners
• political factors including risks	• internal coalitions
• cultural and social change	• norms, values, and partnership culture
• technologies	• technologies
• economic factors including:	• resources, economic factors and risks
❑ *regionalisation* ❑ *competition* ❑ *industry clustering* ❑ *embeddedness* ❑ *corporate change* ❑ *industry reinvention* ❑ *investment*	

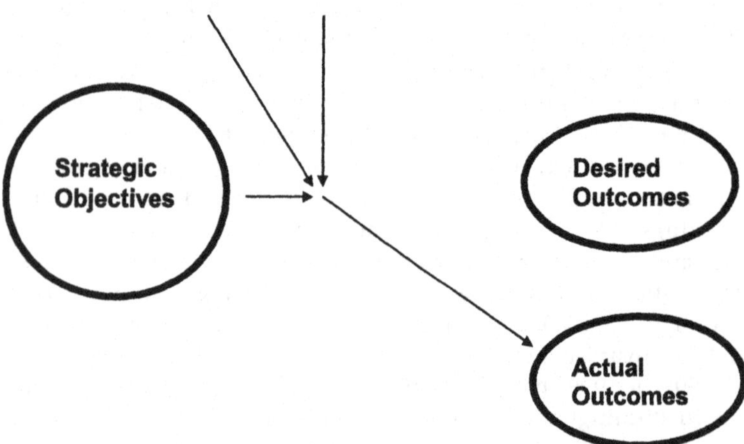

Figure 3.2 Disrupted Rigid Strategy

broader cultural shifts that influence policies, and later chapters refer to the social issues addressed in urban regeneration, but economic change is of fundamental importance. *Regionalisation,* although a longer term process, is itself a major factor because it drives big cities to interact with regional interests that depend upon sustained economic growth. *Regional competition* sharpens vested interests. However, rivalry also forces corporations and cities to work together to try to stay up with competitors by developing regional industrial core competencies. Policy makers regard the *regional clustering* and economic collaboration of corporations with competencies in high technology and other activities as important to sustain economic growth. Also, the *embeddedness* of transnational corporations in dense networks of government and private organisations in regions is important to corporations that establish new production plants and advance their interests through regional networks. *Corporate change* is important especially when it involves company downsizing and process reengineering that can disrupt regional and local economies. Corporate restructuring often accompanies the wholesale *reinvention of industries* when companies are affected by changes in production processes and when high technology knowledge based solutions are applied to meet competitive challenges (Pierre, 1998). In addition, net *investment* needs to be sustained if regions are to prosper, so public and private organisations promote inward investment thus boosting their economic prospects by linking to regional economies. These factors, discussed in more detail below, mean that the health and retention of regional companies are critically important to the well-being of regional economies.

Regionalisation

Rivalry between cities and regions for resources and investment results as companies operate worldwide, but the process of regional change complicates strategy making. The process of the regionalisation of production and its impact on urban change has been viewed in different ways, with Castells (1977) arguing that important productive activities concentrate at the regional level as production organises beyond city boundaries and companies trade outside cities. In his later writings, Castells (1989) refers to the economic uncertainties created by the complex patterns of capital flows between the global and national levels. The information technology revolution has challenged traditional notions of 'local' and created vastly expanded opportunities for international corporations to network their activities across urban and regional boundaries. The local and the global levels increasingly interact as the functions of state institutions divide between different levels and organisations. In the new computer networked society, organisational changes interact 'with the diffusion of information technology' and create new configurations of technology and structure in corporations and regional economies (Castells, 1996).

Harvey (1988) considers the role of space in the organisation of production and argues that urban structure is not simply fashioned. Varied patterns of economic and social activity produce complex changes in the relationships between cities, regional and international economies. Therefore, 'the distinctiveness of the city form, once so apparent as a geographical phenomenon, disappears' (Harvey, 1988, p. 232). Nevertheless, cities support economic circulation, mobilise interests, and generate economic resources for investment in production and commerce. Capital clusters in cities, with corporate headquarters and financial markets located in central business districts, and this is an important characteristic of the modern global economy. This accounts for why 'the economy is hierarchically ordered with local centers dominating local hinterlands' and 'important metropolitan centers dominating lesser cities' (Harvey, 1988, p. 262). However, Harvey down-plays the concrete importance of place in capitalist production by locating cities as amorphous points of reference in a world of global financial flows and 'hyperspace' (Amin and Thrift, 1995, p. 9). Certainly, traditional spatial categorisations in changing economies are inadequate (Massey, 1994) because different spatial levels of organisation do not clearly line up with governmental boundaries, but modern capitalism creates important 'place-based movements' (Amin and Thrift, 1995, p. 9) and produces the conditions for regionally based political conflicts and competition. The different settings for this political competition are usefully identified as falling within what Scott (1998) alludes to as the global, plurinational, national and regional levels. The global is constituted of 'rapidly materialising networks of economic activity' linking corporations, financial institutions and governments. The plurinational groups nations into blocks, such as the European Union, and other trading groups. The national economy is at the level of the nation state while the regional level is 'a vibrant but also extremely puzzling articulation of modern economic and political life' (Scott, 1998, p. 10). Indeed, regions are puzzling to policy makers because of their diversity and their intricate relationships with the other 'levels' both of which make them difficult to think about strategically.

In the European Union, policy makers recognise the implications of these kinds of changes in practical terms. The Brussels Forum on the European Spatial Development Perspective has underlined the complexities of modern regional economies that are functionally integrating local activities into regional economies. It is a message that both the British and the Dutch have acknowledged. The Brussels Forum underlined the need for a balanced, polycentric, urban system that will have to be effectively managed economically to grow and to achieve environmental sustainability (European Commission, 1998). This calls for the integration of economic development and spatial policies better to take account of the connections between communities, cities and regions. Policies will work across borders and will involve alliances between public and private sector organisations with interests in the Europe-wide economy. As the 1998 European Union Urban Forum indicated, these changes will require new approaches to developing local self-governance as policies develop stronger links between productivity and employment and social and economic growth.

Regional competition

Despite the calls for cooperation, competition within and between regions is underlined by diversity of interests. These strengthen disagreements about appropriate definitions of 'the region' and about how national–local arrangements are appropriate for effective regional policy (Keating, 1991). The European Union Nomenclature of Territorial Units for Statistics (NUTS) identifies regional entities, not by challenging national definitions, but by placing national approaches into a general hierarchical framework. According to the NUTS approach, national and local government boundaries are important because they define statutory areas within which local governments derive their authority and where their operational activities are located. However, regions may include several important cities of regional significance that contend for resources. The Randstad encompasses Amsterdam, Utrecht, Rotterdam and The Hague, and defines an area encompassing all those city regions, but the cities pursue their own interests and compete nationally for resources. The English West Midlands is an official regional entity with Birmingham at the centre, but different agencies define their service coverage by different boundaries and towns and cities in the region have widely different interests. Southwestern Pennsylvania is also defined in various ways by different agencies and partnerships that work to different policy agendas. It is therefore difficult to create effective democratically accountable regional institutions despite the focus of economic activity and corporate investment in regions (Le Gales and Lequesne, 1998). Consequently, regional strategies remain problematic for both governments and corporations that, on the one hand, seek cooperation and consensus, but which on the other compete to survive.

Corporate investors are not necessarily loyal to one region, and they are 'acutely aware of infrastructure and human resource endowments when taking their decisions about where to locate, placing both features at the top of their list of requirements' (European Commission, 1994, p. 11). In the USA many local governments market their attractions effectively, and few cities have severely deficient economic infrastructures and social inequalities of the kind experienced in the European periphery. On the contrary, large American cities provide specialist technological and economic support services and corporations benefit from a nationwide distribution system that serves a national and open market for goods and services. The USA is a huge 'single market' where distributive scale economies have aided market expansion and fostered regional expansion. Shifts in the location of economic activity within the USA have transformed the economies of formally lagging regions as in the south where cities such as Dallas and Houston have attracted foreign transnational corporations seeking profitable investment opportunities. New jobs come with the construction of high-technology production plants situated on green field sites and cities such as Atlanta and Denver have developed new commercial and knowledge based industrial competencies as their economies have expanded. Economic competition in the USA therefore produces significant changes within, and shifts between, regional economies. The unevenness of development and the relative decline of some cities serves to underline the inequalities that accompany the transformation process. Some state and local governments have been swept along with the tide of expansion, while other cities, including Pittsburgh, have been forced to adopt more defensive strategies as well as being proactive. Large regionally based corporations have innovated, but smaller enterprises have also played vital roles, as in Silicon Valley in California with its successful high-technology industries. New companies are important to successful and sustained economic development and the regional clustering of firms is conducive to innovation and the incubation of new ideas. The consequence is intense interregional competition with big cities marketing themselves to rival the opposition for inward corporate investment. Cities adopt 'boosterist' strategies, which for Jessop (1997, p. 37), welds regional cities to entrepreneurial objectives that stress growth, competitiveness and internationalisation.

Regional clustering

The quest for corporate and regional competitiveness has a strong political dimension (Wadley, 1986, p. 39) that derives from rivalries between regional economies. Intense competition between local governments for inward investment concentrates policies on the needs of international corporations seeking to locate new production, distribution, and commercial facilities. Dunford and Kafkalas (1992) point to new corporate geographies that link global corporations to localities where the interest driven global companies join public–private partnerships and bring about corporate alliances. An important corporate geography is that of clustering, and Michael Porter's research into clustering has been an important inspiration for many policy makers in Pittsburgh, Birmingham, and Rotterdam. Porter is frequently cited by public officials and corporate executives as having made an valuable contribution to thinking about strategy. Porter (1980) related industrial structure to competitiveness by showing that the quest for competitive advantage had an impact on upon corporate organisation. Porter (1985) argued that competitiveness influenced the activities, performance and innovativeness of firms as well as their effective implementation of corporate policies. If corporate performance was poor, then the corporation would lack the competitive edge in the market. Porter explained how companies became competitive through their critical examination of generic competitive strategies, namely cost leadership, differentiation and focus. Usually, firms had to choose between these generic strategies rather than combining them. Thus, a firm could seek cost leadership in an industry or it could try to provide a unique product or service through differentiation. Focused companies concentrated on industry segments and developed strategies specifically suited to them. This could lead a company to specialise in producing a particular product or service. However, some industries were in decline, others presented major growth potential, and others were in transition to maturity. Companies therefore had to adopt strategies that enabled them to improve their performance to stay ahead, and they could do this by creating value for their customers and controlling their costs. But, this imposed severe strains on companies that did not have a market cost advantage. Porter therefore argued that the firms that succeeded were those that successfully 'positioned' themselves in the market by developing an appropriate generic strategy and culture. They were also the companies that made alliances with others to maximise cost advantages and add maximum customer value. For regional cities and national governments, Porter's analysis influenced the view that the clustering of related industrial and service activities was inherently advantageous and provided favourable conditions for sustained economic growth. Corporations searched for profitable opportunities to invest and they tried to achieve economies of location. This encouraged clustering, as in the expansion of the semiconductor industry in California's Silicon Valley that appeared to provide confirmation of highly effective synergies between new companies that were innovative and profitable. City officials in Pittsburgh thus seek to develop the regional competencies in robotics and high technology knowledge based applications while Birmingham promotes itself as a major auto industry centre, and Rotterdam publicises World Port activities in trading, shipping and commerce. The clustering concept may or may not be economically viable, but it does support the contention of Peck and Tickell (1994) that cities are searching for new institutional and policy 'fixes' that will make them more competitive. Clustering, and other models, are used to justify the development of new technology centers, business parks, local–regional economic linkages, industrial incentives, training schemes and so on.

Embeddedness

Clustering involves the concentration of industrial activities, and embeddedness is about the interlinking of corporate and public organisations in the regional economy. Young, Hood, and Peters (1994) explore the relationship between corporate investment and regional economic restructuring. High technology is a priority as companies construct new production facilities and clusters of companies develop local alliances. Regional partnerships foster local–global coordination, as in Pittsburgh, where powerful private interests have strengthened the competitive culture. Dicken (1992) shows that transnational corporations disrupt the traditional spatial distribution of economic activity because they are able to shift resources and operations between locations (Dicken, 1992, p. 47). Dicken shows that modern production involves the coordination of complex sets of interrelated activities which involve individual companies operating locally to take advantage of particular skills and specialisations. Transnational corporations deploy production operations in different global, national, regional and local locations where they enter strategic alliances with other companies, as in the automobile industry because collaborative ventures and international subcontracting add to the flexibility of production. Amin and Thrift (1995) argue that the global roles of corporations do not necessarily challenge local autonomy because globally active corporations integrate into the local economy and provide economic diversity and capacity. For Amin and Thrift (1995), regional institutions and partnerships therefore assist the process of local integration and the retention of companies. Corporations and local governments develop global strategies that confirm the 'borderless geographies' of production (Amin and Thrift, 1995, p.5), but the global economy is still structured through spatially defined communities that produce a 'global–local nexus' (Amin and Thrift, 1995, p. 5). Globalisation represents 'a greater tying-in and subjugation of localities (cities and regions) to the global forces' (Amin and Thrift, 1995, p. 8) that are reshaping institutions and effecting changes in public and private sector policy strategies. Amin and Thrift (1995, p. 9) thus 'emphasize the continued salience of "place" as a setting for social and economic existence, and for forging identities, struggles, and strategies of both a local and global nature'. The 'local' is therefore a terrain where social interactions occur, conflicts arise, and economic activity takes place. This implies 'place-based' and 'sometimes place-bound' ways of 'living in the global' (Amin and Thrift, 1995, p. 9) generating numerous contradictions inherent in the political, economic and social relationships established in the local.

Corporate change

Corporate restructuring disrupts relationships and creates conflicts between social groups as when companies threaten to relocate or close production plants. In the 1980s, Piore and Sable (1984) explained corporate restructuring influenced by change in the capitalist mode of economic regulation. For them, existing institutions failed to secure an effective match between the production and consumption of goods and this disrupted the 'institutional circuits' (Piore and Sable, 1984, p. 4) that linked production and consumption as regulatory mechanisms. The large corporation 'was the solution to the organizational problems created by the rise of mass production technology' (Piore and Sable, 1984, p. 73), but the institutions that once created prosperity now produced unemployment, inflation and social unrest. Institutional arrangements had failed to keep pace with technological change, and the response to the crisis was a move away from mass production towards flexible specialisation and service sector growth.

Competition between corporations stimulates changes in company organisation, the competitive deployment of information technology and the introduction of new industrial processes. International corporations operate in a world of global communications and networked financial markets where technological change forces organisational adaptation and major structural reforms in both the public and private sectors. By the mid-1990s, writers viewed the strategic positioning of companies and their internal re-engineering of administrative and management processes (Gouillart and Kelly, 1995). Hammer and Champy (1993) provided an account of corporate re-engineering that contributed to the debate in the mid-1990s about industrial innovation. Hammer and Stanton (1995) referred to corporate re-engineering as a necessary aspect of success in competitive in global markets. Global competition forced companies to be aware of the needs of their customers, and new strategies attuned organisational processes to the demands of the market. Corporations moved away from the traditional concerns of planning and control towards an emphasis upon organisational flexibility, innovation, quality service and cost advantage. New corporate strategies involved adaptability in a changing environment, product and process-led innovation, and the provision of high quality goods and services. For Hammer and Stanton, re-engineering was the 'fundamental rethinking and radical redesign of business processes to bring about dramatic improvements in performance'. Re-engineering was about 'making quantum leaps in performance' and about 'achieving breakthroughs' in terms of 'reduced costs, increased speed and greater accuracy' (Hammer and Stanton, 1995, p. 3). Re-engineering for corporate executives was about 'throwing it away and starting over; beginning with the proverbial clean slate and reinventing how you do your work' (Hammer and Stanton, 1995, p. 4). Process re-engineering involved radical changes to groups of related tasks that, taken together, created value for the customer.

Hammer and Stanton's re-engineering went beyond the 1980s-style restructuring or cost cutting within companies. Re-engineering was a fundamental reorganisation of processes implementing a 'radical new principle: that the design of work must be based not on hierarchical management and the specialization of labor but on end-to-end processes and the creation of value for the customer' (Hammer and Stanton, 1995, p. 11). Moreover, re-engineering produced a vocabulary that found its way into the public sector. It was easy to translate the language of the re-engineering corporation for competitive regions, so regional economies could not afford to stand still because others would overtake unless there were radical changes through institutional reinvention and innovation. The trouble was that governments did not adapt as fast as the most dynamic corporations, but enthusiastic public officials often tried to short-circuit the government system and work in informal networks that were more conducive to getting results, especially working with companies facing global competition (Peck and Tickell, 1994, p. 281).

Reinventing industries

Corporate re-engineering was induced by international competition and because many of the 'older' American and European corporations that had dominated the expansive phase of capitalism between the 1950s and 1970s were confronted by competitors in the East Asian economies and Japan (World Bank, 1993). Davidow and Malone (1992, p. 24) refer to the deteriorating position of American companies in the 1980s as competitors produced new products and services and grew fast with high profits. The customisation of products and quick delivery made American companies 'lose their edges' in a world of continuous change (Davidow and Malone, 1992, p. 2). Hamel and Prahalad (1994) showed that between 1980 and 1989, many transnational corporations expanded their workforces, but productivity failed to match the pace. They refer to corporations including BASF, Data General, Westinghouse, Borden, Dresser and General Motors that all reduced their establishments by 'downsizing'. They list Monsanto, Union Carbide, IBM, Digital and Kodak that also shed substantial parts of their workforces (Hamel and Prahalad, 1994, p. 7). They faced stagnant markets, declining profits, and falling market share. Corporate transformation followed as unemployment and decline in whole industries became major political issues. Downsizing, according to Hamel and Prahalad, played an important part in rationalising companies and cutting out waste and inefficiency, but companies often set no clear limits to the process that resulted in many companies making short term decisions rather than ones that would sustain long-term growth. In this way, downsizing undermined strategic direction, but innovative companies went beyond restructuring by re-engineering. Lean production combined with company-wide policies on quality enabled American and European companies to emulate the successes of the Japanese. Many American companies became involved in a process of catching-up rather than in a drive to regain the competitive leading edge in global markets. For Hamel and Prahalad (1994, p. 15) it was 'not enough for a company to get smaller and better and faster, as important as these tasks may be; a company must also be capable of fundamentally reconceiving itself, of regenerating its core strategies, and of reinventing its industry'. The companies that failed fundamentally to reinvent their core strategies were, according to Hamel and Prahalad, the ones that most needed to reskill their workforces, sell off unprofitable parts of their businesses, and restructure their operations. The lesson was that the future successful competitive companies had to go beyond the re-engineering of processes to whole industry reinvention.

The early 1990s ushered in a period of economic recession in major developed economies followed by an upturn in the USA in the late 1990s and economic crisis in East Asia. Companies accelerated their restructuring going beyond re-engineering to diversify, decentralise, specialise and expand. The economic conditions facing companies changed as the emphasis shifted away from downsizing to corporate repositioning in important markets. The downsizing of the 1980s indicated the complex problems facing the 'first generation' of transnational corporations that had originated largely in Western Europe and the USA. The new generation of transnational corporations in East Asia provided competition on quality and production, but the older corporations were able, even by the late 1980s, to emulate the competition. The relative cost and productivity advantages between innovative corporations became less significant as indicators of disparity between competitors. Hamel and Prahalad suggest that companies therefore reinvented whole industries to capture increased market share, and in the 1990s, this extended beyond mere cost and product competitiveness. Companies identified new products and new corporate competencies as the whole customer interface needed to be reconfigured if companies were to stay ahead (Hamel and Prahalad, 1994, p. 73). Managers needed to look beyond the narrow confines of existing markets as old demarcations between national corporate entities were being broken and companies collaborated internationally. For Hamel and Prahalad, this brought a new 'strategic architecture' that rested on the new core competencies and the reduction of product development times. Global branding did not restrict companies to making products suited to specific national markets, but allowed them to create internationally recognisable brands and product ranges.

Investment

While large corporations already located in particular regions undergo these substantial changes, regions are also influenced by external capital flows and foreign investment. Ietto-Gillies (1992) refers to Vernon (1966) who stressed the importance of technological gaps between nations after the Second World War. In nations with a higher development of technology, there was a propensity for corporations to invest away from their home bases. The USA had a developing technological base so it could exploit overseas markets by selling high quality goods to consumers that were increasingly keen to spend their disposable incomes. The domestic production of commodities in the USA thus expanded production overseas as products became more standardised internationally. Ietto-Gillies detects a neglect in the literature of the cumulative advantages bestowed by technology upon local communities because writers have concentrated too much on the demand factors attracting investment from overseas. Some recognised the importance of foreign direct investment in host economies and that the investment decisions of international companies were strategic in helping companies to compete with others in world markets. Ietto-Gillies shows that more recent theories of Foreign Direct Investment (FDI) focus on the need for transnational corporations to adapt production to world markets by subdividing production processes and placing labor intensive production abroad often through subsidiaries or alliances.

The USA, the European Union and Japan account for most of the top transnational corporations and, despite crisis in the former 'tiger' economies, many East Asian countries are heavily committed in Europe and the USA. Young *et al.* (1994) stress the impact of corporate investment locally and show that regional economies are affected by joint ventures, acquisitions and alliances that come high on company agendas. The economic recession in the European Union and the USA during the early 1990s encouraged the trend to companies concentrating on corporate strategy and restructuring, and the economic downturn interrupted the earlier boom in FDI, most notably from Japan. Nevertheless, Young *et al.* (1994) refer to the efficiency of production arising from FDI in the international economy that sustains comparative advantage. This demonstrates a potential economic benefit to regions and localities arising from the operations of transnational corporations especially in economies that suffer as a result of a lack of investment and growth.

The health and retention of regional companies

The above factors, taken together, crystallise to influence the overall health of regional economies and they produce multiple problems and opportunities for strategy makers. Global capitalism expands the horizons of big corporations, but still leaves them, workforces and domestic regional economies vulnerable. Corporate re-engineering, international capital flows, and company alliances focus regional economic development strategies on promoting the interests of companies within regions. For city governments, there is a political role to play in networking with companies to improve employment conditions and create a favourable investment climate. However, in the economic sphere local governments often have to be more responsive than proactive and this can reduce their immediate room for manoeuver, but as networkers they are often crucially important to corporations seeking to expand their market opportunities.

The Pittsburgh region displays all the consequences of the multiple factors associated with economic restructuring. Pittsburgh is headquarters for several large transnational corporations that control substantial resources within the USA and globally. These corporations have all undergone profound changes through company restructuring, downsizing and corporate re-engineering. Major transnational corporations in Southwestern Pennsylvania include high technology companies, major manufacturers, financial and service companies. Eight 1996 Fortune 500 largest United States corporations had their headquarters in Pittsburgh. These were the USX Steel Corporation, Alcoa, Westinghouse Electric, H. J. Heinz, PPG Industries, PNC Bank Corporation, Mellon Bank Corporation and Consolidated Natural Gas. The USX Corporation, Alcoa, and Westinghouse featured in the Fortune 500 listing of the world's top corporations. The USX Corporation headquarters in Pittsburgh has long been regarded as symbolic of the formerly mighty steel industry that contributed to Pittsburgh's industrial strength in the nineteenth and early twentieth centuries. However, Lubove (1996) provides a vivid account of the region's steel industry and its dramatic decline as the industry restructured in the 1980s. He shows that the counties around Pittsburgh were once home to the powerful steel manufacturers, but these areas suffered most when steel declined. By 1900, the region already over-specialised in coal, iron, steel, glass and electrical machinery production, and these mass production heavy industries created the conditions for future economic crisis and the downturn of the regional economy. Lubove describes the industrial landscape of smokestack industries that prevailed into the 1960s and 1970s. The US Steel Corporation Donora Works in Washington County closed in 1966. Manufacturing decline accelerated leading to the severe 1980s shakeout of steel jobs as US Steel and other companies closed obsolete plants. Lubove (1996, p. 7) lists the closure of US Steel plants at Rankin, Duquesne, Clairton, Homstead, and McKeesport, and Jones and Laughlin (LTV Steel) closed plants at South Side and Hazelwood and Wheeling–Pittsburgh closed its works at Monessen in the 1980s.

The USX illustrates that traditionally large regional corporations expanded their interests and redefined their operations globally in order to respond to changing markets. The USX Corporation in the 1990s, through its Marithon Group, was involved in wordwide marketing, exploration, and production with interests in oil, natural gas and refining. The US Steel Group produced steel and steel mill products, managed mineral resources and provided engineering and consultancy services. Other USX businesses included real estate and financial services (Grubb and Ellis Company, 1995, p. 92). Alcoa, the worlds leading producer of aluminum, operated globally in twenty-six countries making products used in the automotive, construction and aerospace industries. The company was involved in bauxite mining, smelting, shipping, recycling, ingot fabrication and industrial research. A Pittsburgh corporate review indicated that the corporation's products are used 'in every important market worldwide' (Grubb and Ellis Company, 1995, p. 6). Also in Pittsburgh, the H. J. Heinz Company is one of the world's largest food manufacturers. Heinz enjoys extensive market penetration with well-known brand products in Europe, North America, Australasia and Asia, with sales outside the USA accounting for forty-three per cent of the company's $8 billion annual sales (Grubb and Ellis Company, 1995, p. 46).

During the early twentieth century, the Westinghouse Electric Corporation became a major force in world electronic and electrical product markets. By the1990s, Westinghouse was a multi-industry corporation with extensive media, transportation and nuclear waste management interests (Grubb and Ellis Company, 1995, p. 94). By 1996, its media activities included the ownership of television stations, a joint venture with CBS, radio stations and a production company. The global involvement of Westinghouse forced substantial restructuring as the corporation attempted to reposition within changing markets. The reorientation of the company through 'demerger' in 1996 reflected its increased global orientation and its eventual shift away from its Pennsylvania base as it expanded into media. In 1997, the new CBS Corporation united the media interests of the Westinghouse Electric Corporation and the former CBS. This corporate reinvention followed Westinghouse Electric's earlier agreement to sell its power generation business to Siemans and to divest its remaining industrial businesses from the media company. The management team for the spin-off decided to phase out the Pittsburgh Gateway headquarters which would be downgraded as a regional outpost of CBS in New York.

In Rotterdam too, the accent is on catering for the interests of transnational corporations and preserving and developing the city's existing economic base. The Rotterdam City Development Corporation (OBR) claims that the city is a first class location for international corporations. 'The presence of Unilever, Shell, Robeco Group, NedLloyd, and Nationale Nederlanden, to name but a few, is proof that Rotterdam delivers the goods' (OBR, circa 1996, p. 6). The OBR headlines the attractiveness of the city to international investors and the development of the central business district with its high rise buildings, like the imposing Nationale Nederlanden building completed in 1993. In the central business district 'many of the top name service companies have established offices here primarily due to the location and the phenomenon that the character of a company expresses itself in the design and choice of its architecture' (OBR, circa 1996, p. 7). The OBR thus portrays Rotterdam as the natural home for transnational head offices like Unilever N.V. For the OBR, it is an international city where the central business district bustles with all the excitement of the most go-ahead modern cities. Rotterdam, according to the OBR, offers access to international trade centres, conference facilities, the World Port, hotels and top quality leisure facilities. It is the image of the corporate city, the place to be and to be seen. It is a place where 'offices are taking an ever more prominent place' and where 'international lawyers, insurance companies, chemical groups, and banks' (OBR, circa 1996, p. 10) have congregated. The boosterist appeal continues by reference to the city as a place where business can network and where new business and industry development nodes such as Delta 2000–08 at the port, the Noordrand project to the north of the city, and the Kop van Zuid project in the centre provide profitable investment opportunities.

Despite the outward confidence, city politicians and officials in Rotterdam remained concerned about the overall competitive position of the city in the mid 1990s. Global conditions increased the uncertainties about the longer term role of the World Port with Antwerp to the south and other ports adding to intense competition in world distribution and transportation. Migration from abroad increased fears about the social stability of a city with large communities of disadvantaged ethnic minorities. Even companies such as Royal Dutch-Shell, the international oil group that had long had a major presence in the city, could no longer automatically be assumed to be sure bets for future expansion. Royal Dutch announced in September 1998 that its Dutch national head office in Rotterdam was to close as part of a worldwide assessment of the group's operations. Group chairman, Mark Moody Stuart, speaking in San Francisco (*Financial Times*, 19 September 1998) referred to the worsening business climate for the group and the need for the group to take radical measures to reassert its market position. The closure of the national offices in the Netherlands, France and Germany marked the start of a campaign to rid the group of entrenched interests resistant to change, and the start of a major transformation that would affect all its employees.

Birmingham projects its industry, but the increasingly branch plant nature of the local economy means that the attention of city policy also is on internationalising the city by attracting leisure and service activities, financial institutions and organising high profile sports and international conference events. Big companies with production in the city include Bass Taverns, GEC Alsthom Mero-Cammell, GKN Hardy Spicer, IMI Titanium, Lucas, and SP Tyres. According to the Birmingham Economic Information Centre (1996), the city was still Britain's largest manufacturing centre in terms of Gross Value Added and it had the second largest number (800) of medium sized and above company head offices, compared with other

British cities, outside London. These do not rank with Pittsburgh's Fortune 500 corporations and headquarters, but they are important in the British economy.

Economic restructuring, including in the nearby Black Country, attracted banks and service businesses into Birmingham's central business district during the 1980s, and the city stimulated development with new shopping centres, an international convention centre, and a national sports arena. However, Birmingham City Council is aware of the difficulty of retaining companies that restructure and relocate. The Rover automobile plant at Longbridge in Birmingham was recently bought by BMW of Germany, and Jaguar cars by Ford. Volume automobile manufacturing in the city therefore now rests in foreign ownership, one response to which has been the refinement of a regional strategy that envisages new industrial activities. However, following the acquisition of Rover by BMW, the city council was confronted in late 1998 by the threat of BMW to close the Rover plant at Longbridge. This prospect, brought about by low labour productivity, threatened plant closure or a substantial reduction in jobs. The labour unions agreed to new working practices, and BMW eventually agreed to continue to invest in Longbridge but with a reduced labour force and promised public funding. Such problems led the city council to concentrate more on high technology as a way forward for Birmingham in the future. Initiatives like the high tech Millennium Point project near the central business district could provide a boost for new technologies and help change the direction of the local economy to overcome the decline of the city's former staple industries, but the uncertain situation at Rover symbolised the often fragile confidence in a region increasingly vulnerable to decisions taken in distant boardrooms.

The need for strategic flexibility

The above issues underline the need for effective strategies in economic development. Collaborative strategy making enhances the political role of local governments even though they are challenged economically. The British Urban Regeneration Association (1997) underscores the necessary combination of ingredients for effective local and regional strategies under the headings of connections, clusters, capital and community. Policy makers network to exchange information and learn lessons from other local governments and corporations about these issues. Company clusters must exploit the opportunities of pooling resources for economic development and work with local governments to bring this about. Mayer (1992) argues that local governments can therefore influence the process of economic change and influence economic restructuring by working with corporations to produce economic infrastructures conducive to growth. However, Mayer recognises that governments frequently have to respond to circumstances largely beyond their control. The case studies that follow show that big city governments can adopt proactive and effective regional economic development policies, and in the USA and Europe, cities have learned from private sector approaches (Halachmi, 1995; Chisholm, 1997). In Pittsburgh, Birmingham and Rotterdam public officials have thus pursued policies that relate to the market conditions mentioned above. In all three cities, there is a distinction between regional and local strategy that reflects the unevenness of economic development and spatial organisation, although in Rotterdam 'the region' is more closely identified with the contiguous metropolitan area. Figure 3.3 summarises the nature of regionalisation, political competition, administrative fragmentation and types of strategy adopted in each of the three urban regions.

In all three cities, the public–private partnerships within the urban regions, have addressed the environmental changes discussed above through innovative strategies. Government intervention and strategic planning that once provided a model were replaced by public facilitation and greater strategic manoeuverability. However, formal strategy often leads officials and politicians to misread the external environmental and rely on self-fulfilling prophecies and top–down bureaucracy. A more flexible and intuitive approach to strategy, particularly in situations where complicated situations are frequently likely to be very risky and unpredictable, becomes a necessity. Elements of this flexibility are evident in the three cities to greater or lesser degrees. In Pittsburgh, the city government combines with strong business leadership and a public policy commitment to a market oriented approach to urban and regional economic development. In Birmingham, regional and city strategies reflect a move away from hierarchical planning that favours market policies and greater business involvement in the local and regional economies. In Rotterdam, the political climate has changed with greater stress on public–private partnership and new flexible roles and structures for public institutions. The emphasis in the three cities has tended to be upon urban revitalisation, measures to stem urban

Figure 3.3 Strategy in Three Cities

Aspects of Strategy and Politics	Rotterdam	Pittsburgh	Birmingham
1. Regional and city strategies.	1. The Randstad constitutes a wider region, but metropolitan area of Rotterdam defines scope of city's regional strategic approach. Various attempts to reform government.	1. Business led regional strategy for Pittsburgh urban region, and a city economic development strategy with increasing collaboration with county.	1. A regional strategy for the West Midlands, and a Birmingham economic development strategy for the city. New regional bodies in 1999.
2. Pattern of politics.	2. Emphasis on restrained competition. Joint actions develop through consultation, political consensus, and community group involvement.	2. Political competition, especially at regional level, makes partnership necessary. The local level involves community groups working with business groups with stress on 'empowerment'.	2. Political competition at local and regional levels. Shared visions sought, but emphasis on centrally funded programmes with community groups collaborating at the local level.
3. Administrative fragmentation and central influence over strategy.	3. City council retains important controls over economic development projects. Collaboration with provincial government and centre. Central government decentralises decision making to city.	3. Fragmented regional organisational environment with substantial role for business leadership groups. Little federal government influence over local and regional strategy making process.	3. Fragmentation with some semi-autonomous agencies, but central influence exerted through networks and partnerships. Strong role of central government restrains networks. Reorganisation of regional administration.
4. The style of strategy making for region and city.	4. Tradition of planning in city but economic development strategy more market-oriented. Use of scenarios. City attempts to be competitive through inward investment and strong economy. Decentralised decision making.	4. Entrepreneurial approach with city having less of a role in strong business-led regional strategy. City and regional strategies aim to overcome decline of traditional industrial base with aim of strong competitive market economy.	4. Strong emphasis on planning has changed to greater market concern, but retention of city and central government involvement. City and regional organisations defend manufacturing base and seek inward investment.

poverty and support for company retention. The three cases that follow therefore display distinctive approaches to strategy making where cities have adopted measures to meet new competitive challenges, but where strategy makers are constantly revising strategies better to relate to changing conditions.

Policy makers today have strategic awareness, but that does not ensure success in an aggressively competitive world. The most successful city governments have to respond quickly as economies restructure and companies reinvent their core businesses. They need effectively to combine intuition with considered analysis of situations (Mintzberg, 1994; Kickert and Koppenjan, 1997) without formalising their approach as rigid planning. However, many of the strategies that follow have quite tightly defined aims and objectives that invariably need revision in the light of reality and external events (Kouzmin and Jarman, 1989). Flexible strategy, therefore, relying more on intuition, is less common in city and regional strategic thinking than perhaps in the private sector. The rigidities contained in vision statements, plans of action and specific desired outcomes are still common hallmarks of public sector strategy making. The kind of synergy that characterised the business environment in California's Silicon Valley provides an illustration of the sort of culture that can develop, while cities that are unable to attract the best companies are increasingly vulnerable to the vagaries of the market, deskilling of local workforces and continued structural economic weaknesses. Moreover, if a gulf between the region and local communities occurs, citizens are isolated from influential regional networks and deprived of economic opportunities. Community groups in Pittsburgh, Birmingham, and Rotterdam are active at the local level, but until recently they have had only limited influence in regional economic development partnerships (Imbroscio, 1997). Each case shows the closer relationships that are developing between communities and public–private sector partnerships as policy makers draw connections between the global and local interests of their regions and recognise the importance to corporations of more highly skilled workforces.

4
Strategy and Partnership in the Pittsburgh Region

This chapter, and chapter five, are based on research carried out between late 1996 and late 1998. This chapter shows that in Southwestern Pennsylvania the external economic and political environment of local government has produced a variety of strategic public policy responses and a complex interorganisational landscape. Chapter five covers urban regeneration and community involvement in local economic development in Pittsburgh and surrounding areas.

Southwestern Pennsylvania is a relatively compact region by American standards, where governments, corporations and public agencies all have their own policy strategies and where administrative fragmentation makes it difficult for policy makers to arrive at a consensus about what should be done. The problem of administrative fragmentation has been so dysfunctional that corporate leaders have encouraged a shared regional vision to improve economic competitiveness through public–private partnerships and innovative policies. Coherent regional strategies have sometimes been difficult to produce because local political interests have hampered efforts to produce a broader regional perspective. Local politicians have often tried to please electors by resisting tax raising measures supported by local and state interests to boost economic development, and corporate leaders have had to overcome the frustrations resulting from poorly coordinated development efforts. Also, a shared regional vision has been difficult to achieve when public and private organisations have adhered to their own interpretations of what the 'region' and regional identity mean. The different interests and constituencies of regional organisations in the Pittsburgh region have thus strengthened political competition despite the functionality of those differences for the individual groups involved.

Regional strategies and complex governance

In the United States, regional strategies have to account for fragmented regional governance where state and local governments and public agencies work with reference to their respective operational territories. Bradshaw (1988) refers to governments and businesses in the USA during the 1980s developing different regional perspectives in economic and social policies. However, traditional definitions of regions as geographically defined jurisdictions do not hold because almost any regional boundaries can be contested (Bradshaw, 1988, p. 7). Social networks, economic activity, and high technology processes cut across boundaries and invalidate notions of regions as neat spatial or socioeconomic units (see chapter three). Bradshaw's 'elements' of 'regional difference' magnify the problems for strategy makers. He describes the differences between regions and the lack of agreement over how spatial units contribute to regional variation. For Bradshaw, the USA, in contrast to other nations, appears remarkably homogeneous, but it is also a 'nation of geographical variety,' with variations in natural resources, cultures, perceptions, and political orientation. Historical factors underpin such differences. The nation therefore has a 'mosaic' of regions (Bradshaw, 1988, p. 53) that depend upon the interaction of economic, social, demographic, political and cultural influences within American society. Consequently, regionally based public and private organisations reflect this diversity and they each respond differently to structural economic changes. Bradshaw refers to the recognised census regions, but many federal government agencies have used these as territorial designations while others have not. Public programmes often use other designations. For example, in 1965 the federal Economic Development Act established the Economic Development Administration that defined economic development districts by clustering groups of counties within states.

In Southwestern Pennsylvania, the problem of defining a region concerns the different territorial interests of public agencies, county and city governments. The importance of the region around Pittsburgh was underlined when business and government recognised the need for a broader policy focus. The Allegheny Conference on Community Development (ACCD) was an important private sector-led partnership that resulted from the Allegheny Conference on post-war planning organised in 1943. The ACCD was formally incorporated with an executive committee in 1944, and it championed the interests of Pittsburgh within the region. The ACCD and other private sector organisations and local governments have produced definitions of the region that have reflected the changing market conditions and contingencies confronting different organisations. Research for a 1993 'white paper' on regional economic development, produced for ACCD, showed the Pittsburgh–Southwestern Pennsylvania region as five, nine or thirteen adjacent counties (ACCD, 1993, p. 4). The white paper defined a Pittsburgh metropolitan region as the five counties of Allegheny, Beaver, Fayette, Washington and Westmoreland. In the white paper, a thirteen county region would add Armstrong, Bedford, Butler, Cambria, Greene, Indiana, Lawrence and Somerset counties. A recently revised census definition usefully refers to the city at the heart of a wider Metropolitan Statistical Area (MSA). Within the MSA, Pittsburgh lies within Allegheny County that has 130 municipalities within a 731 square mile area. The six county MSA includes Allegheny, Beaver, Butler, Fayette, Washington and Westmoreland counties (see Map 4.1). Pittsburgh increasingly cooperates with Allegheny in economic development and planning, and other local government bodies and public agencies run various city services. Indeed, the rationalisation of city–county governance has been a major concern for the corporate sector as well as the City of Pittsburgh because of the economic interrelationships linking the city and Allegheny County. The MSA definition therefore provides a convenient and appropriate definition of the 'close' economic metropolitan region with Pittsburgh as the central city. The MSA includes important related economic activities of industries and services in the counties and it provides a useful basis for measuring social and demographic changes. For example, Figure 4.1 provides recent Bureau of the Census population estimates for the MSA.

However, public and private organisations would be constrained if they used the MSA as an exclusive regional definition. The Southwestern Pennsylvania Regional Planning Commission defines Allegheny County and adjacent counties including Armstrong because they are relevant to its specific planning focus. The Penn's Southwest Association, and other major regional economic development groupings, define a region as the ten counties of Greater Pittsburgh (Map 4.1) as this marks a wider economic entity that partners find useful operationally. The Pennsylvania Economy League defines a region as the twenty-nine counties of western Pennsylvania because it portrays a significant economic market within

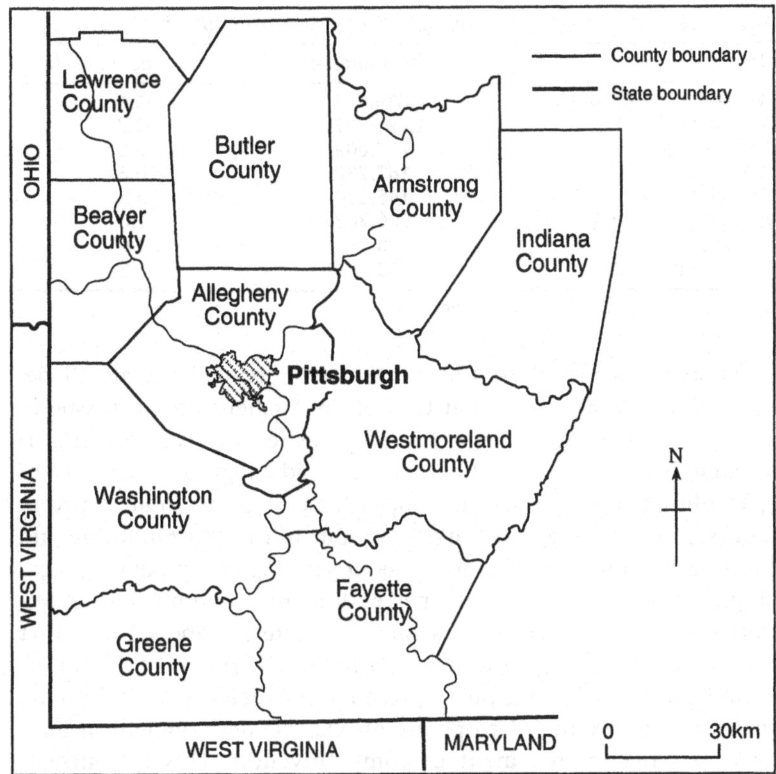

Map 4.1 Greater Pittsburg: A Current Definition

Source: The Pittsburgh Regional Alliance, 1999, showing recently included Indiana County.

the domestic economy of the USA. The various definitions reinforce the business (ACCD, 1992, p. 1) view that economic activities cut across local government jurisdictions, and so giving too much importance to arriving at a single definition of the region therefore misses the point. The importance of different definitions is that they reflect various organisational perspectives and interests in the regional economy. Differentiation can undermine regional coherence and challenge democratic accountability in government, but it also enables individual agencies and private organisations to relate to specific market conditions by using territorial definitions relevant to their individual needs.

Policy-relevant conceptions of 'region' develop from ideas about

Figure 4.1 Estimates of the Population of the Pittsburgh MSA, 1990–6

County	1996 estimate	% change 1990–6
Whole Pittsburgh MSA	2,379,411	–0.6
Allegheny County	1,296,037	–0.3
Beaver County	187,009	–0.5
Butler County	167,732	10.3
Fayette County	145,628	0.2
Washington County	206,708	1.0
Westmoreland County	376,297	1.6
City of Pittsburgh	350,363	–5.3

Source: US Bureau of the Census, 1997.

space and territory even though territorial boundaries are ill defined. Soja (1996) claims that territory is a salient political issue in a very particular way. It is not territory as legal jurisdiction that is so significant, but it is the social and ideological constructs of spaciality that are important. Soja's 'first-space' is about territory and 'second-space' about the ideas about territory. Combining and transcending these, third-space is about social groups producing ideas about territory that fuse with perceptions of economic and organisational interests. Territory is symbolic of interests and values rather than being exclusively a rational economic definition of spatial organisation. So 'the region' represents the defence of economic interests, the making of corporate profits, the development of markets, and the improvement of competitiveness. This discourse is echoed in the political and interorganisational competition in regions, with globalisation adding an international dimension to regional identity reflecting the politicisation of space. Responses to economic environmental change are thus cast in the language of competition, the need for shared visions, and the value of cooperation and community.

State level initiatives

In the USA, the discourse is shared between substate level regional organisations and the states themselves. The states in the USA are major economic players and they are political organising spaces for corporations and regional alliances that defend strong regional interests. The Commonwealth of Pennsylvania would easily rank as a major region in Europe measured by economic activity and land area. Public–private economic development partnerships in Pennsylvania organise to expand the global market presence of regionally based corporations, and state and local governments try to boost the competitive regional economy. In Pennsylvania, the state government provides a range of incentives to promote economic development through support for public–private sector alliances and technology networks with the Governor's Action Team bringing together economic development professionals to provide a one-stop economic development office to help business and to foster partnerships. By 1996, the team, based in the state capital at Harrisburg, had branch offices in Pennsylvania and overseas offices in Brussels, Ontario, Frankfurt and Tokyo. The team linked corporations with state agencies, utility providers, local economic development offices, development companies and financial institutions, and it marketed Pennsylvania internationally as a location for investment. It encouraged foreign inward investment and backed joint ventures involving organisations based in Pennsylvania. The team worked with corporations to identify industrial sites and alerted firms to a range of financial incentives to increase investment and support industrial retention. Incentives included low interest loans for land acquisition, and grants for infrastructure improvements and workforce training all of which were matters of high priority for Governor Tom Ridge.

Partnership and vision

What drives regional cooperation when there are so many agencies producing visions and policies for the region and when such differences are often highly functional? It is clear that Southwestern Pennsylvania's tradition of business leadership is contextualised by the interplay of the corporate private sector with public agencies. In the Pittsburgh region, a motive behind cooperation between the private and public sectors is precisely the existence of regional diversity so that the public and private sectors have come together to overcome overlapping policies and to rationalise the economic development process. Sbragia (1996) reinforces the view of fragmented governance in the USA where cities, agencies and corporations jointly adopt entrepreneurial strategies to improve regional competitiveness. For Sbragia, cities are investors and states are entrepreneurs, and this implies that they increasingly behave like corporations. However, there is often a tension between cities and states even when there is cooperation to promote regional objectives. However, for Dodge (1996) regional policy in the USA involves the joint activities of community leaders, governments and citizens in the organisation of a range of activities to address commonly perceived policy problems. For Dodge, regional strategies mobilise the efforts of 'lower tiers' of government and communities working together to deal with increasingly difficult problems that require collaborative strategies. New approaches to regional governance require state and local governments to define new relationships with federal government, overcome administrative fragmentation and ensure political accountability. Dodge maintains that this process is leading to a substantial modification of existing approaches to economic development and community regeneration because state and local governments implement strategies that link the local, regional and global objectives.

Dodge refers to contending models of regional governance affected by these changes. According to Dodge, the Balkanization model of regionalism produces a scatter pattern of poorly coordinated local governments. Under such circumstances, local governments are unlikely to cooperate unless there are compelling reasons to do so, but in spite of the reluctance of some local governments to work together, economic and demographic changes force the issue of cooperation onto the political agenda. Dodge points to another form of regionalism that for city governments implies a clearly defined hierarchy of service delivery responsibilities between local governments with a strong metropolitan authority at the center. In practice, elected representatives have been reluctant to challenge either of these models for fear of either undermining local autonomy or enhancing the powers of large metropolitan governments at the expense of smaller suburban or rural ones. However, Dodge claims that old ideas about regional governance are breaking down as economic change brings innovation and new strategic policies. The American 'regional renaissance' (Dodge, 1996, p. 46) has the potential to empower communities as they link with the region and as political leaders recognise the opportunities and benefits of new approaches to commonly perceived problems.

Regional politics therefore combines both cooperation and inter and intra regional political competition. In Southwestern Pennsylvania, despite the success of some partnerships in formulating regional strategies, intraregional differences between state, county, and local governments persist. These differences involve a range of policy issues, but in economic development they often reduce to some very basic considerations about such matters as tax and local consultation. Local politicians are aware of the unpopularity among the electorate of tax raising measures, and their views about regional issues tend to derive from their desire to represent local constituents. County governments have regional interests as well as local ones, and that reveals a difference between their regional and local policy orientations. The situation is similar for big urban regions. Pittsburgh is important as an economic hub, so the city council has representation on all the major regional economic development partnerships, but it also defends its own interests. The city is politically important within formal and informal policy networks, linking with the counties, public agencies, and corporations and it plays a role within major business led regional consortia. Pittsburgh has economic significance, and city departments provide professional advice and help to facilitate large development projects. The city council and the powerful specialised Urban Redevelopment Authority of Pittsburgh (URA), act as conduits between the corporate private sector and public authorities, and city and URA officials network to establish business-like relationships between the public and private sectors. Regional and local economic development in Southwestern Pennsylvania thus depends upon strong leadership from the corporate private sector coordinated with state, county, local government, and public agencies. Public agencies employ skilled officials who have developed expert roles within networks and partnerships. These are people with an intimate knowledge of local markets and the economic development process.

Business leadership and strategy

Business leadership in economic development goes back a long way in the Pittsburgh region. By the 1930s, Pittsburgh was known as 'the smoky city' where environmental pollution was the consequence of rapid and ramshackle industrial development. The problems were so acute that business leaders were concerned that the good health of the city's workforce was at stake. The business community believed that new policies could improve conditions and produce gains in industrial productivity. The Pennsylvania Economy League had researched the problems of the state and regional economies, and there was a recognition that local governments and corporations had common economic interests. In the 1940s, the Pennsylvania Economy League and the Pittsburgh Regional Planning Association sought to bring agencies together to develop regional economic opportunities, and business leaders and politicians created a powerful partnership known as the Allegheny Conference on Community

Development (ACCD) to deal with the city's problems. The line-up included major business leaders, Pittsburgh's Mayor David Lawrence, Wallace Richards (Secretary of the Pittsburgh Regional Planning Association), and Robert Doherty (President of the Carnegie Institute of Technology). It was ACCD that provided the inspiration for partnership and public–private cooperation through the 'renaissance' of the central business district of Pittsburgh, the improvement of communities, and the development of regional initiatives.

This role established ACCD as a major player in economic development and urban regeneration during the post Second World War years. However, the intense global competition and the decline of key markets in the 1980s took a heavy toll on the region's staple industries. In 1993, the ACCD commissioned an important report on the competitiveness of the greater Pittsburgh region. A committee under the leadership of Robert Mehrabian, then President of Carnegie Mellon University, produced a 'white paper' on regional economic cooperation. The white paper focused upon the economic strengths of the region's economy, but it also revealed disturbing weaknesses that needed to be addressed by adopting new mechanisms and an effective strategic regional policy. Mehrabian outlined the historical development of public–private partnerships in economic development in Southwestern Pennsylvania by reference to a layer cake of organisations. Figure 4.2 depicts the layer cake showing how partnerships developed since the end of the Second World War by increasingly broadening their policy horizons and focusing their activities regionally. Post war growth stimulated organisations and networks with specialised roles relevant at particular times, as for rebuilding Pittsburgh's central business district and shedding the city's poor public image. Many of the partnerships had structures that resembled 'starburst' or 'spider's web' networks of multiple interests that were highly adaptable to changing conditions (Quinn, Anderson and Finkelstein, 1996, pp. 357–7). This allowed many partnerships to pool the resources of their members to develop their specialist roles and functions in local economic development. The white paper showed that the developmental process led to new organisations during the 1960s era of urban crisis in American cities with partnerships concerned with urban revitalisation and community development. There was therefore a functional basis for new organisations, but those functions changed over time as the region came to be perceived as a more important economic factor influencing the economic fortunes of Pittsburgh. The ACCD, for example,

Figure 4.2 Layer Cake Partnerships

Era of change	Partnerships
The Future Era	The development of regional partnerships
Economic Crisis Era	Mon Valley Initiative Allegheny Policy Council Southeastern PA. Growth Alliance Southwest PA. Industrial Resources Center Pittsburgh High Technology Council Community development corporations Pittsburgh Partnership for Neighborhood Development Steel Valley Authority
Urban Crisis Era	Action Housing Model Cities/Vista Perry Hilltop Citizens' Council Homestead Economic Revitalization Corp. People's Oakland Community Technical Assistance Center
Postwar Growth Era	Allegheny Conference on Community Development Regional Industrial Development Corp. Urban Redevelopment Authority of Pittsburgh Smaller Manufacturers' Council Chamber of Commerce Penn's Southwest

Source: Based upon presentation in the White Paper (ACCD, 1993).

expanded its regional vision and developed new organisational capabilities as its membership grew and as its reputation as an important business-led player strengthened. The ACCD created a strong organising core within a starburst-style network that provided a clear mission for business leaders in urban revitalisation. By the 1990s, the developmental process had resulted in a complex and fragmented organisational map with its many agencies and private sector organisations with overlapping roles and jurisdictions. Often there were conflicts and bureaupolitical turf fights between public agencies. Partnership characteristics varied, being influenced by the transformations in the regional economy that produced quick responses to crises, adapted structures and politically expedient solutions. The layer cake was therefore suggestive of the incremental nature of the development of many organisations and their structural transformations over time. Some, like ACCD, had developed highly professional operational capabilities with strong strategic apexes and specialist staff functions while others were more localised and informal.

The white paper referred to the lack of coordination between regional organisations that meant that the task for business was to weld closer links and develop a clear strategy for Southwestern Pennsylvania. The Regional Economic Revitalisation Initiative (RERI) that followed the white paper supported research and consultation through Carnegie Mellon University's Center for Economic Development. The initiative marked the creation of a community-wide strategic consultation process to deal with the issues raised in the white paper. Public and private sector partners agreed that 'members of the community did not have a shared vision of what the region's economy could be: 'A crucial ingredient, consensus, was missing' (RERI, 1994, p. 5). The 1994 RERI report thus proposed a process to advance to a situation where all groups, including local community organisations, could agree on common policy objectives. Organisations spread across the region produced 'the first concerted planning process for the Greater Pittsburgh region in a decade' (RERI, 1994, p. 5) and a report entitled 'The Greater Pittsburgh Region: Working Together to Compete Globally' detailed a phased approach to building regional partnerships. Phase one concerned the development of a shared regional vision. The vision depended on consensus, so a Leadership Consortium brought together representatives of corporate, local government, county, labour, education, community and religious organisations. The consultation process involved local communities through town meetings and the KDKA–TV 'Decision Point' programme so that 'every facet of our diverse community has been involved in understanding our problems and opportunities and developing a strategic plan and flagship initiatives to achieve our economic vision of working together to compete globally' (RERI, 1994, p. 6). Phase two involved the organisation of work groups to identify the central problems and opportunities in regional economic revitalisation. Consultation and discussion informed the development of projects and facilitated extensive networking between the organisations involved in the process. The main themes covered by the work groups covered the siting of industrial activities and offices, business expansion, the identification of core business activities, and business startups (RERI, 1994), and the partners produced a series of studies on regional innovation, competitiveness, high technology, and the promotion of the regional economy. The studies covered regional economic performance since 1990, a comparative assessment of Southwestern Pennsylvania as a business location, and how to retain and expand manufacturing in the region. There was also discussion of how partners could revision the region by building upon Southwestern Pennsylvania's manufacturing and technology capacities. The reports thus indicated the seriousness of the commitment of business to a strategic effort and the desire of partners to research the competitive position of corporations and the prospects for the regional labour force. Phase three was about the development of specific projects and ways to create 100,000 jobs in the region. A work group on demonstration projects carried out an assessment of proposals in high technology, bioengineering, advanced manufacturing and environmental industries. A work group dealing with 'destination development' viewed the opportunities in the region to develop tourism, and another group studied the possibility of linking regional economic development organisations to expand business opportunities. Other groups viewed the problems of the infrastructure, labour policies, and tax and regulatory issues.

Working together – or still competing?

The culmination of the activities of the work groups was the formulation of a strategy based upon recognition of key policy challenges that were contained in a regional strategy document for a newly established Working Together Consortium (WTC, 1996) under the chairmanship of Richard P. Simmons. Simmons, as President of ACCD and Chairman of Allegheny Ludlum, had exceptionally good connections and was qualified to get together diverse regional interests. His ACCD pedigree ensured that WTC would be well placed for networking with the City of Pittsburgh, the URA, Allegheny County economic development officials, and more widely. It also meant that he could attract a high calibre leadership team to guide the consortium. The WTC established a broadly based consortium-partnership of public and private organisations committed to mobilising the resources of the Pittsburgh region and capitalising upon the region's core economic competencies (WTC, 1996, p. 2). Figure 4.3 shows some of the partners that came together in the WTC and illustrates the strong representation of major industrial and financial players in the regional economy. Many of these were to use the WTC as a vehicle to lobby local governments and state agencies.

Richard Simmons stressed the overarching nature of the WTC and the importance of his leadership team in coordinating policies. The RERI report had referred to a 'blueprint' for regional economic development that superficially had connotations of formal planning and the centralised monitoring of policies. However, if the strategy was to work through partnership and public–private action, there

Figure 4.3 The Working Together Consortium

Examples of Working Together partners.
The following is a sample of the members of the Working Together Consortium in 1996:

Chair, Richard Simmons, Allegheny Ludlum Corporation and President of the Allegheny Conference on Community Development
With members from:
Alcoa
Allegheny Ludlum Corporation
Carnegie Mellon University
City of Pittsburgh
Deloitte and Touche
Duquesne Light Company
Duquesne University
The Enterprise Corporation of Pittsburgh
Heinz, USA
IBM
Mellon Bank
Minority Enterprise Corporation of Southwestern PA.
Pittsburgh High Technology Council
Pittsburgh Public Schools
PNC Bank Corporation
PPG Industries
Price Waterhouse
Sony Electronics
Southwestern PA. Industrial Resource Center
University of Pittsburgh
Westinghouse Electric Corporation

William Coyne, US Congress
Rick Santorum, US Senate
Richard Cowell; Thomas A. Michlovic; Michael R. Veon, Pennsylvania House of Representatives

Plus representatives from educational, health, religious, business, training and other groups of organisations.

Source: WTC, 1996.

could be no conflict between private initiative and regional visioning. The RERI report therefore explicitly referred to a 'clear strategy' that should not impose a top–down process of regional development. If the 'blueprint' appeared to be a somewhat inappropriate description for a process that was supposedly to be attuned to changing market conditions it nevertheless provided the impression of action and coherence. It also represented the desire by the partners to retain control over the partnership's development and strategic policy commitments.

The region had to overcome its economic weaknesses, loss of

population, and lack of competitiveness by building on its strengths, and there had to be a revival of 'entrepreneurial vitality' (WTC, 1996, p. 1) with an economic climate conducive to growth in the twenty-first century. Many corporate and political leaders perceived relative economic slippage in the Pittsburgh region when compared with others and they expressed concerns about other cities, such as Cleveland, overtaking Pittsburgh as a model for partnership. The WTC strategy was based on the creation of an economic development hub for the region, flagship projects, the expansion of tourism, and education and training initiatives linked to high technology. The strategy developed themes that related to the repositioning of the region within both the domestic and global markets, and it expressed the concerns of government and corporate partners who were aware of the weaknesses of the region and the loss of talented professionals and business persons to other parts of the USA. Many of the strategic commitments, such as 'reviving' entrepreneurial vitality, therefore tended to reflect the worries of many politicians and business people in the region, and the partners were aware of the need to implement the aims of the partnership through a regional consensus that would order political competition through the common identity of otherwise contending interests.

The commitment to 'build on the strengths that hold the key to future job growth' (WTC, 1996, p. 1) was typical of strategies adopted in other regions that had lost manufacturing capacity. The partners maintained that the region had core competencies that had to be developed to create the basis of future wealth and employment. Research found that five major industrial strengths contributed to the region's potential global strength. Metalworking was identified as an important cluster accounting for forty per cent of the region's manufacturing employment. Other key clusters were chemicals and plastics, health and biomedical systems, information and communication products and services, and environmental products and services. Public and private sector initiatives would ensure the development of the clusters, the establishment of new research and development opportunities, and the involvement of local communities. In innovative fields, such as robotics, biomedical applications, environment and Maglev production, the region could lead the world. The region itself could become a huge technology park (WTC, 1996) in order to meet the challenge of global competition, but only if there was a recognition of the development of new patterns of economic activity and organisation. The partners wanted to invest in wealth and job creation. Investment was the key to the development of infrastructure and human resources, and the challenge could be met through partnership such as between the City of Pittsburgh, Allegheny County and the ACCD. This link-up would provide a funding strategy to build business parks and stimulate employment. However, while there was little discussion of the capacities of regional organisations and their potential for meeting such substantial objectives, there was a recognition of the need to overcome resource constraints and take actions appropriate for dealing with the problems facing the region.

The regional partners wanted to create an 'economic climate for the twenty-first century' which depended on a broad regional interpretation of partnership and, in effect, the achievement of regional restraint in political competition between groups. The idea was that success in economic development depended upon good industrial relations, shared visions of change, a favorable tax regime, and boosterist projects. It was significant that a management–labour agreement on industrial relations would underpin the economic development strategy because it was important for labour to adapt to new technologies and flexible work processes through partnerships with management to help the clustered industries. The labour–capital partnership approach also covered housing, transportation and the completion of various infrastructure projects. The economic climate had to be supportive of enterprise and initiative to achieve the ambitious objectives set. The capital–labour agreement would thus create conditions that would help to 'establish a new spirit of teamwork in economic development' (WTC, 1996, p. 19). The labour–capital accord was crucial to the meeting of the partners' objectives because employment and business growth implied that policies would involve local communities. This meant that women, minorities and dislocated workers should have a stake in economic development, and that they should have access to business advice, startup funds and training. Mark Nordenberg, Chancellor of the University of Pittsburgh, later defined the connection between training in communities and the WTC employment strategy when he claimed in a study for WTC that 'workforce development equals economic development' (Fletcher, 1998, p. 20).

The RERI report claimed that the 'Greater Pittsburgh region set the standard for public–private partnerships.' Partners had to 'set a new standard for partnership by creating a framework for collaboration to meet the challenges we face today, (RERI, 1994, p. 19).

No single agency could 'deliver the services and strategies our region requires. Borrowing from the lessons of the business revolution under way, we must create concrete mechanisms for the ventures and alliances that can deliver comprehensive marketing and retention strategies' (RERI, 1994, p. 19). The labour unions and other organisations would play a part in working with companies to provide training and workplace experience for young workers and students. This point later linked to the Clinton Administration's desire to foster connections between local communities and growing regional economies where local people could play a part in expanding industries and services. 'Building one economy' was therefore about forging local–regional linkages. Pittsburgh could not 'be a world class region of two economies, one of opportunity and quality jobs and one without. The imperative of building an economy which brings opportunity to all citizens and communities has emerged as a priority in almost every meeting and outreach session conducted to shape this blueprint' (RERI, 1994, p. 23).

The WTC viewed the organisational and governmental fragmentation in the region as a serious fetter on the future development of the region economically. The large number of public and private sector organisations with interests in the prosperity of the region had, according to the consortium, to come together more effectively to achieve economic success. This produced a management coordination problem for business leaders and other partners in the WTC. The problem was how to create the effective organisation of a partnership consisting of a large number of members each with different interests and ideas about the future. The preferred approach was to provide an organisational structure for the partnership that built upon the specialist capacities of the partners through a decentralised structure, but it was also important to acknowledging the legitimate demands of member organisations by creating an effective structure of authority and delegation. The WTC itself therefore represented organisations that contributed to the objectives of the consortium while retaining their individual autonomy. The WTC theoretically ensured that all major partner organisations had a role to play in the regional strategy. However, the WTC depended upon the ability to define a regional capability through a strategic apex that had the task of coordinating the efforts of different organisations, and this sometimes proved difficult. The strategic apex consisted of a leadership team that encouraged the development of the specialist organisations delivering particular aspects of the WTC strategy and these organisations reflected strong corporate interests. For example, the Southwestern Pennsylvania Growth Alliance partnership dealt with industrial site reuse and issues arising from the reclamation of contaminated land. The growth alliance featured strongly in ACCD strategy because of its innovative approach to multiagency collaboration and its role in promoting regional interests through lobbying the state and federal governments. The Pittsburgh Regional Alliance (PRA) resulted from the creation of the WTC with members including the influential Greater Pittsburgh Chamber of Commerce, the Penn's Southwest Association, the Regional Industrial Development Corporation, the World Trade Center of Pittsburgh, and the Pittsburgh High Technology Council. The alliance promoted business cooperation through the provision of a one stop hub to cater for the needs of businesses seeking assistance with corporate development, location, trading, marketing and other issues. The PRA developed extensive links with other partnerships and a strong relationship with the Minority Enterprise Corporation connecting to local communities. The PRA also campaigned on a range of issues concerning local business taxes, Allegheny County government reform, and federal policies affecting business (Greater Pittsburgh Chamber of Commerce, 1998, p. 19).

The PRA approach was successful, but partnerships still proved hard to pull together, and public discussion about an alleged lack of strategic consensus in the region was covered in the Pittsburgh media. For example Thomas McConomy, as chairman of the WTC committee responsible for implementing its theme of establishing a 'new spirit of teamwork,' stated in the Pittsburgh Post-Gazette (20 December 1998) that there was still too much fragmentation in regional economic development and that some individuals in leading organisations distrusted one another when making key decisions. It was an opinion that echoed the Mehrabian white paper's earlier observation that groups in the region did not always work harmoniously.

New regionalism: more partnerships

Partnerships that operate beyond the confines of traditional local government indicate broad corporate horizons and a strengthening regional focus for policy. In simple terms, the Pittsburgh urban region is important to corporate leaders because a citywide focus is too narrow to relate to the special problems of the competitive regional economy. The development of such perspectives by business leaders in other regions in Pennsylvania shows a growing meso level focus in the strategic thinking of business partnerships. In Northeastern Pennsylvania for example, public and private sector partners have developed a strong regional council in economic development that represents a determination to meet the challenges posed by new patterns of economic development, market change, and the problems created by mobile capital. The regional council concept arises from the desire of cities, towns and counties across the USA jointly to address issues that affect them and which do not restrict strategic policy making to narrowly defined geographical or jurisdictional areas (Dodge, 1996; Economic Development Council of Northeastern Pennsylvania, 1992, 1993; Grossman, 1994a, 1994b). In areas with substantial fragmentation of government, as in the Pittsburgh region, there exist subcounty councils of neighbouring jurisdictions (Dodge, 1996) that have stimulated job training, business enterprise and the integration of women and minorities into the workforce. Franzini, Smith, and Frakt (1994) show that new partnerships among private enterprise, local communities and different levels of government enable the alignment of community groups with the private sector to support pro-market visions assuming that communities gain through the development of strong local enterprises. Initiatives in housing, business development, job creation, crime and education provide 'long-term multifaceted' solutions (Franzini *et al.*, 1994, p. 24) where community and corporate interests intersect.

Policies in the Pittsburgh region generally rely upon the recognition of the common interests of the public and private sectors, although business organisations have lobbied hard against tax measures and regulations proposed by state and local governments. The private sector has an interest in promoting the region to improve economic growth, to improve the competitiveness of Pittsburgh companies in the wider economy and to revision the city and region in the service and tourism markets. But the realisation of genuine regional consensus involving corporate and community interests remains problematic. With senior business leaders such as Thomas McConomy publically expressing concerns about the lack of consensus and trust in regional economic development, there seems some way to go before an effective shared commitment can be consolidated. This applies to communities, and while local people do link to the PRA for example, direct engagement with local people in regional partnerships is not yet as extensively developed as locally. The Pittsburgh case shows that partners recognise the need for regional coordination to overcome the duplication of resources and policies, but this recognition by itself does not mean that partnerships necessarily produce the desired outcomes. Partnerships in Southwestern Pennsylvania allow different organisations to pursue their own specific objectives and specialist services while at the same time agreeing to cooperate through the WTC consortium and other alliances, but this is different to the 1940s model of corporate involvement focused around influential business 'barons' with their single-mindedness to get things done.

5 Pittsburgh: Partnership and Community Empowerment

This chapter shows that Pittsburgh's business and political leaders have involved neighbourhood groups closely in programmes and that local partnerships have increasingly developed policies that are important for regional economic growth. However, by the 1990s, the once cohesive local corporate elite in Pittsburgh was less homogeneous, being represented by active executives rather than the old corporate barons of the 1940s. The chapter shows that these modern business leaders worked with professionally competent public officials to improve competitiveness in an economy where local communities were encouraged to engage with the regionally active businesses by developing new skills and capacities. While there were many successful public–private partnerships, it was essential that collaborative programmes developed a synergy through effective joint working and focused community leadership. In the search for effectiveness, the discourse about 'empowerment' was about capacity-building through the acquisition of community development and business skills. Empowered communities contributed to the wider economy, took advantage of the business and social opportunities that resulted from collaborative working and benefited from improvements to the urban fabric resulting from economic growth and job creation in local industries.

Linking local communities to the region

The Reagan administration encouraged states and big city governments to work with business to enhance competitiveness through less reliance on federal government. The public sector borrowed management techniques from the private sector to run community services (Bledsoe, 1993) and local community leaders worked more effectively with business leadership groups. According to Reagan, community empowerment developed links with business and expanded local self reliance. By the late 1990s, according to Vice President Al Gore, 'businesslike' federal government behaved more like the private sector and was more efficient, oriented to the customer and less encumbered by wasteful bureaucracy (Gore, 1997). For example, city governments facilitated property deals and worked with companies to relocate production and provide business services (Kresl and Gappert, 1995; Savitch and Vogel, 1996).

Al Gore's 1997 seminar series, 'Community 2020: A New Future for the American City' identified the need for effective policies for sustainable development, crime reduction, educational betterment and improved race relations. An important seminar theme dealt with cities as important regional hubs where public–private partnerships contributed to economic competitiveness and community empowerment in regional economies. The discussion about the future of the cities during the Clinton Administration's second term reassessed urban issues, but many local economic development practitioners complained that policies lacked coordination in the absence of a coherent approach to federal urban policy. Against this scepticism, Community 2020 dealt with the role of metropolitan regions in the new global high technology economy. Community 2020 adopted a broad perspective that envisaged an urban policy spanning outwards to the wider global and regional economies. It suggested that coordinated regional policies could help to overcome local community problems such as unemployment and poor housing. Despite the criticisms, it was a view that represented a significant theoretical rehabilitation of 'the city' and a practical recognition that cities would play an important part in sustaining the nation's competitiveness.

The US Department of Housing and Urban Development (HUD) report, *America's New Economy* (HUD, 1996), showed that improved regional competitiveness was important to the success of local economic development. According to the report, 'metropolitan regions' were centrally important in an era of information technology where new and diverse industry clusters were overcoming traditional single industry dependencies. The HUD report advocated closer links between business and government for competitive strategies, but competition also intensified rivalries between the big cities. In *The State of the Cities* report (HUD, 1997) the problems for older cities remained with their urban ghettos and worn-out infrastructures and congested central business districts. As the report showed, ethnic minority involvement in urban affairs had expanded, but the urban poor still demanded access to a policy process that often seemed to be remote and impersonal (Glazer, 1994; Hunt, 1997). Communities suffering high unemployment, social disadvantage and communal tensions were encouraged by HUD to be more self-sufficient through neighbourhood enterprise (Taub, 1994), but groups still lacked influence over key decisions. These issues were relevant in Pittsburgh communities that suffered from decay and neglect in the wake of economic decline. According to a study by Pittsburgh University Center for Social and Urban Research (1994), this was a city where 25.9 per cent of the population was black and where the incidence of poverty in ethnic minority communities was among the worst in the USA. The enterprising culture in Pittsburgh had for a long time fostered political cooperation to deal with the problems of urban decay, poverty and poor housing, but such deeply-rooted problems had persisted into the 1990s, and in many ways appeared to be worsening.

Business leadership

After the Second World War, Pittsburgh's top business leaders instigated major urban regeneration initiatives that created a new mood of confidence in the city. The physical expression of the renewal effort provided one of the most enduring and impressive symbols of urban revival in the USA. Pittsburgh's new skyscraper skyline came to signify an era of civic progress and provided visible evidence that the city could reinvent itself in spite of the stigma of decline. Lubove (1996, p. 72) suggests that 'Renaissance One' in the 1940s and 1950s prevented Pittsburgh from becoming one of the problem cities of the 1960s by escaping the dire consequences of urban decline that befell cities such as Gary and Newark. The Allegheny Conference on Community Development (ACCD) spearheaded Renaissance One that led to the dramatic high-rise development of the downtown area. Mayor David Lawrence delivered a new 'consensus on community policy' (Lubove, 1995, p. 111) that had as its aim smoke control, flood control and the development of the central business district. For Stewman and Tarr (1982) Renaissance One marked the highpoint of cooperation between the business and political elites in the city. They argue that public–private partnerships became deeply woven into the political and economic fabric of the city in the nineteenth century and this continued into the twentieth century when civic leaders rationalised local government and improved the image and condition of Pittsburgh. Renaissance One thus achieved public–private collaboration and the efforts of notable Republican corporate leaders with the support of Mayor David Lawrence's Democratic political machine (Stewman and Tarr, 1982). Lubove shows how the business–city coalition affected all levels of the city government. Each objective required the use of public funds and the commitment of the city to administrative and policy support and, for Lubove (1995, p. 111), the coalition established a 'reverse welfare state' where public funds supported corporate private sector ambitions. All this was reminiscent of the existence of a powerful business 'growth machine' backed by the city council (Logan and Molotch, 1987; Harding, 1995; Ferman, 1996). However, the political alliance between business and the Democratic political machine forged a strong relationship between the public and private sectors that was later challenged and disrupted. Stewman and Tarr (1982) argue that public–private partnerships changed during 'urban renewal' in the 1970s under a new political leadership that lent less support to the corporate sector. Mayor Peter Flaherty, elected in 1969, stood on an 'I'm nobody's boy' platform (Stewman and Tarr, 1982, p. 89) that distanced him from labour unions and corporations. For Stewman and Tarr, this produced a different political environment for the private sector with the Mayor favouring community interests as opposed to those representing the central business district. Stewman and Tarr refer to the transformation in city government departmental leadership and the refocusing of collaborative activity at the middle management levels in the city and corporations. Working relationships between the public and private sectors were now 'not only at the top, but also at the staff level' (Stewman and Tarr, 1982). Relations between Flaherty and the ACCD deteriorated as their agendas diverged and as the Mayor became less interested in big redevelopment projects and more keen to cut city government spending. Nevertheless, under Flaherty the ACCD remained active and managed to reposition itself as an advocate for development projects and a valuable partner in downtown development. A new Mayor, Richard Caliguiri, supported 'another Renaissance' (Stewman and Tarr, 1982, p. 94) but business leadership was no longer focused around a core of notables as it had been during Renaissance One. Without a political machine, the city assumed a more managerial role in the development process and developed an expertise as a development facilitator and 'planner and initiator' of public and private initiatives (Stewman and Tarr, 1982, p. 95). Renaissance Two occurred from 1977 when the second major wave of urban redevelopment got underway and led to the expansion of the Urban Redevelopment Authority of Pittsburgh (URA). This public authority influenced the success of Renaissance One and it extended its ambit over housing and community issues during the Flaherty years. Lubove thus portrays the city council and public authorities fulfilling valuable roles for the private sector, and using public funds that enabled business leaders to pursue civic transformation.

In the 1990s, Democratic Mayor Tom Murphy's urban regeneration and economic development depended upon the city council as an important strategic agent in a market-driven development process. Through the Office of the Mayor, the city produced a strategic development approach that envisaged a long-term involvement for the city through strong political leadership. Murphy argued that cities

> that once relied on external funding to define economic development opportunities today face the realization that economic opportunity – growth, development, and jobs – no longer comes preassembled. State and federal governments that once provided the lion's share of financing now play a strategic role, filling the gaps and providing leverage to attract private dollars. (City of Pittsburgh, 1996, p. 1)

Murphy detailed a market-influenced vision of a thriving growth-driven city with a city council committed to responding to consumers in the new technological economy. The priorities were the development of the river front, industrial site reuse, downtown development, and the improvement of neighborhoods. The aim was to create an economy responsive to business and the community by taking advantage of new technologies and market opportunities. The city council aimed for a 'customer-first approach' to business attraction and retention, the purchase and assembly of land, and the 'timely application of various development financing and capital formation tools.' (City of Pittsburgh, 1996, p. 1). The city's customer-focused policy included the creation of a team of expert economic development executives working with companies, banks, financial institutions, and state and local government departments. Community organisations could also call upon the team for advice in initiating and funding projects, and the city worked closely with the ACCD where projects had a regional impact.

All these activities, including industrial site development and business assistance, provided a high-profile political role for the city in urban regeneration and community development. The identification with the Mayor and the Deputy Mayor, Tom Cox, with the strategy emphasised its political importance. However, the politicians were involved in a high-risk approach, and Murphy regarded the acquisition of land and property as a vital ingredient in the strategy with a development fund providing support for major projects. Collaboration with business, community organisations, the state, public agencies and Allegheny County underpinned a policy that involved the city as strategy maker, negotiator and partner (City of Pittsburgh, 1996). The commercial risks were substantial, which was especially evident when partners invested in projects that had mainly long-term returns depending for their realisation upon favourable market conditions. For the city, this implied both economic and political risk-taking where the success of projects could produce political rewards. The city gained from success, but its political leadership stood or fell by the outcomes. This market-interventionist strategy depended upon the support of the private sector, but the city could not rely on the 'hidden hand' of market forces, so it was in partnership with corporations and nonprofit organisations that the most effective risk sharing could be achieved.

Public authority and private action

Mayor Tom Murphy's strong reliance on the URA linked the city's vision to public action to underpin partnership with the private sector. The URA was a powerful mechanism for driving the city forward and making it attractive for corporate investors. For Murphy, the URA was a vehicle for promoting the city regionally and combining urban revitalisation with neighborhood improvement. The Mayor-appointed URA Executive Director at the time of this research, Mulugetta Birru, agreed that the investment of the city in its local communities was essential for economic growth, and Murphy and Birru confidently predicted that the URA could achieve its goals through public powers aided by state and federal funding. The URA worked within the Urban Redevelopment Law dating from 1945 and amended by the Commonwealth of Pennsylvania in 1998. The law provided the legal basis for redevelopment authorities in the state and defined the objectives and powers vested in the URA and its status as a public authority. The law detailed the URA role in urban regeneration in blighted communities that were unsafe, unsanitary and overcrowded. The law empowered the URA to deal with planning issues and to overcome problems associated with economic and social disadvantage. It justified a public authority to deal with these issues because private enterprise needed the assistance of the public sector in planning, development control and legal regulation. A public authority could use its special powers to clear derelict sites to prepare them for private investment, assume the power of land acquisition, and initiate site clearance and disposal. The URA could package sites working with the private sector to make land ready for development, and the URA could acquire properties where there was a danger to public health or severe social problems. This extended the URA's remit into housing, especially where living conditions fell below habitable standards. The law also detailed the ways in which the development authority could work with other public and private sector organisations. While the URA was a corporate public body that worked within the authority laid down by the state, it could collaborate with others and 'take on its own additional investigations and recommendations' (Commonwealth of Pennsylvania, 1988, p. 9). This meant that the URA could initiate projects according to need and could cooperate with any government, school district or municipality acting as an agent of the state or federal government. The URA could borrow funds from private lenders and invest funds held in reserves or sinking funds in legitimate schemes and trusts. The law stated that the URA could 'make and execute contracts and other instruments necessary or convenient to the exercise of the powers of the authority;' (Commonwealth of Pennsylvania, 1988, p. 12). The URA could make and repeal bylaws, conduct public hearings on planning issues, administer oaths, issue subpoenas, and designate areas for residential development. The URA was not itself a partnership, but was a facilitator and enabler for partnerships and corporations. In many respects it acted like a core operational and strategic apex for others. It provided a core competency in property and economic development from which others could benefit. Within networks of property developers, the URA frequently acted as a surrogate partnership core; functionally similar to the core of a starburst network (Quinn, Anderson and Finkelstein, 1996, and chapters two and ten). However, unlike classic private sector starburst networks, the external partners did not legally 'own' the URA, despite the original corporate involvement in establishing the authority in the 1940s.

While the URA was a single development authority, its orientation to external organisations was multifaceted and organisationally specialised, and strategy in the late 1990s combined an expanding regional focus with a strong local community focus. The major development projects in the city were of regional significance and they clearly promoted urban regeneration, but they also provided what the URA described as 'solutions for a changing market' (URA, 1996, p. 5). The big development objectives required prioritisation for land acquisition and assembly, the provision of development financing and capital formation, and business attraction and retention. In 1996, the URA identified a series of projects that most met these priorities including a large site at Nine Mile Run, the South Side Works site, the Hays Ammunition Plant and the East Liberty Sears Site. In addition, a variety of other major development sites added 500 acres of land to the URA portfolio. On the development financing and capital financing side, the URA established the Pittsburgh Development Fund and expanded the use of tax increment financing. An important objective was to expand the use of a variety of possible funding sources, including state funds, and the use of tax credits. The URA provided professional services for site selection, development coordination, and business financing and had responsibility for private investment in businesses and business support and for the operation of an Urban Development Fund for development. The URA Housing Department carried out the authority's housing programmes and work associated with the other neighbourhood housing programmes. These activities showed the degree to which the authority had developed a full range of professional services and established a strategic position in the city's real-estate development market.

By the mid 1990s, clearly defined objectives (URA, 1996, 1997) provided a useful picture of the kinds of partnerships developed in local initiatives. Projects had to be commercially viable with clear purposes to ensure that they contributed to the local and regional economies. As with private sector investments, the URA supported projects that packaged different sources of funding through public–private partnerships. For example, the North Shore project, with private and public financing, involved the relocation of the ALCOA corporation from its old downtown building. The project enabled ALCOA to develop an impressive headquarters on a North Shore river-front site close to the recently opened Andy Warhol Museum. The Carnegie Mellon University–NASA Robotics Engineering Consortium initiative was another important URA project. The Robotics Institute of Carnegie Mellon University established a national Robotics Engineering Consortium to exploit the commercial potential robotics technologies. Consortium funding included contributions from federal and state sources and urban revitalisation associated with the project enhanced the Lawrenceville neighbourhood. Over the early period of operation of the project, approximately $50 million worth of funding was expected to support development, research, administration and the operation of the consortium. Such URA projects produced many economic gains for communities even from highly commercial projects. Expected outcomes included the creation of new jobs, training for local people, environment improvements and the stimulation of local economies. The URA also claimed to have improved the city's neighbourhoods with schemes such as East Liberty Gardens that used state, URA and bond finance to rehabilitate a large low income housing project. Neighbourhood organisations welcomed these benefits, but they continued to demand other tangible improvements for communities including action on crime, drugs and other social problems.

The regional orientation was just as important for the URA as the neighbourhood focus. In 1997, the URA and Allegheny County signed a corporation agreement that marked a major shift in focus for the URA to regional partnerships. The URA and Allegheny County agreed to share resources and to work jointly to attract business to the region. The URA, the Redevelopment Authority of Allegheny County and the County Department of Economic Development came under the common directorship of Mulugetta Birru of the URA to achieve the shared vision. The URA continued to recognise the need to carry out its strategy in a professional and commercially aware manner, but now large projects were promoted by county agencies as well. Cooperation extended to major initiatives originally promoted by business groups, for example to create a nationally significant tourist attraction in Pittsburgh, the construction of a new baseball stadium, a new football stadium and an expanded convention centre. By mid-1998, major approvals for funding for these schemes had been obtained with additional support coming from federal and state government sources.

The URA thus worked regionally within a highly competitive real estate development market by adopting an active approach to marketing the projects that it supported. The URA case illustrates the dynamic nature of an organisation that intervened in a changing market in close collaboration with other public and private organisations. Working arrangements, based on commercial and neighbourhood links produced strong collaborative links between the URA and its partners, and in 1998 there was a prospect of a major reconfiguration of inter-organisational structures and relationships and the enhancement of county powers through a home rule charter. The picture was of active public authorities at the city and county levels broadening their regional perspectives and diversifying methods of market intervention.

Community empowerment and restrained politics

The involvement of the URA in neighbourhoods required community involvement in an ordered and restrained policy process. Neighbourhood organisation in Pittsburgh was complicated by the existence of formal and informal networks and local starburst-style partnerships. This produced varied local political competition involving public, private and nonprofit organisations, although direct local community leadership involvement tended still mainly to focus at the local as opposed to the regional level.

Ferman (1996) refers to competition between community and business interests in Pittsburgh in the 1950s that gave way to cooperation and partnership. She analyses groups in relation to the 'growth machine' (Ferman, 1996, p. 17) that charaterised the development-led regime in Pittsburgh after the Second World War. Ferman (1996) and Lubove (1996) trace the developing relationships between business, community groups and the city council arising from the ACCD's drive for citizen support for its early redevelopment efforts. The Allegheny Council for the Improvement of Our Neighborhoods (ACTION Housing Inc.), established by ACCD as a nonprofit organisation, championed business involvement in community revitalisation and helped ACCD overcome community opposition to extensive redevelopment. Ferman (1996) contrasts the experiences of neighbourhood groups in Pittsburgh and Chicago where contrasting political cultures influenced group incorporation. In Pittsburgh, the old political machine 'withdrew from key parts of the policy process' (Ferman 1996, p. 33) whereas in Chicago the machine attempted to defend its popular electoral support by running campaigns against local groups and racial minorities. She finds that in Pittsburgh community organisations established a cooperative working relationship with the city council while in Chicago the record was less encouraging. Chicago politicians historically opposed activist groups and resisted their inclusion into legitimate policy making. Pittsburgh's neighbourhood organisations challenged 'the growth machine' (Ferman, 1996, p. 13), but in doing so they gained access to the city's politicians and public officials and joined local partnerships. Ferman shows that in the 1950s and 1960s Pittsburgh's economic and political 'regime' became responsive to the demands of local organisations, but this partially offset the local power of business elites. Politicians operated through 'the civic arena' (Ferman, 1996, p. 145) whereas in Chicago their base was in the electoral arena. She refers to the establishment of the Pittsburgh Partnership for Neighborhood Development (PPND) in 1988 as a funding body that allowed groups into the city's policy process. The PPND provided a core managerial competence for various groups organising like the starburst arrangement described in chapter two. For Ferman, contested politics in Pittsburgh involved neighbourhood organisations because there was political space for them to represent local demands. She maintains that the incorporation of local groups depended upon a political culture in the city where community access to policy depended largely upon who politicians were and what they were prepared to stand for (Ferman, 1996, p. 16). Pittsburgh's business and political elites thus structured their power in a different way to the political and business elites in Chicago. Significantly, Ferman locates this contrasting structuring of political power in the different institutions that elites established in the two cities. Pittsburgh's leaders expanded and mobilised community resources by using the civic partnership model that extended deep into the local community, but this produced a comparatively elitist incorporation of groups when compared with the varied politics of the electoral arena. In Chicago, poor neighbourhoods were left out of the expansionist developments associated with the new service economy because political patronage was more important than community involvement.

In the 1990s, Pittsburgh city council and business leaders continued to back urban regeneration through community improvement. In both economic and political terms, the elitist model described by Ferman was now strengthened by business sharing risks and responsibilities with community leaders. Neighbourhood partnerships therefore focused upon well-defined objectives and community leaders supported management innovation and organisational effectiveness. Community organisation was no longer the reserve of the gifted amateur, and community professionals became attuned to market needs and the professionalism of nonprofit organisation. Like managers in the private sector, community leaders were accountable for programme outcomes and they raised funds for complex projects and managed large administrative staffs. Community leaders represented neighbourhood interests, but to be successful they had to convey their ideas effectively and needed to assume responsibility for their actions.

The nonprofit community development corporations (CDCs) provided an important stimulus for neighbourhood-based self-help and partnership. The CDCs concentrated upon the improvement of communities through economic and social initiatives and the rehabilitation and construction of modern affordable housing. The CDCs obtained federal funding for projects including day-care provisions, job schemes, and environmental improvements. Local partnerships combined the efforts of banks, corporations, foundations, government and others to 'pool their resources and expertise to effect lasting change in their communities' (Ford Foundation, 1996, p. 1). Access to programmes was not enough because groups also needed to be active and motivated to pass on community development skills to others. The PPND worked with CDCs and received Ford Foundation support in promoting such active initiatives. Ferman (1996) regards the formation of the PPND as a turning point that bonded relationships between business and the community in Pittsburgh. In 1993, Donald Tizel, then Chairman of the Mellon Bank, summarised the PPND role as using the expertise in communities by building partnerships involving a network of CDCs whose members would invest the time to revitalise their communities. It was the public and private sector partners that could 'provide the resources to bring reality to ideas'. Tizel described the role of PPND as that of an intermediary, bringing together people with partners that could enable people to work more effectively. Working with the City of Pittsburgh, the URA, banks, foundations and others, PPND supported a wide variety of programmes through grants and loans. The intermediary role benefited the PPND and the individual CDCs that it supported. The approach maximised the benefits of collaboration through the sharing of business risks, and the PPND encouraged neighbourhood regeneration through decentralised structures and managerial autonomy. The PPND thus worked as a decentralised networked organisation, but with a strong support framework provided by PPND's own staff. By 1996, the PPND strategic apex consisted of a board of directors with representatives from Mellon Bank, Penn's Southwest Association, the URA, the City of Pittsburgh, PNC Bank, local universities and foundations. 'Keystone' partners included the Howard Heinz Endowment, the Pittsburgh Foundation, the Ford Foundation and the Richard King Mellon Foundation. Other partners included Dollar Bank, Landmark Savings Association and the PNC Charitable Trust. The PPND therefore acted as a conduit through which management expertise and funds went to local CDCs. The PPND was a relatively flat organisation, but it provided high value support to local mangers and by 1998 the organisation was flexible enough to allow for expanding local programmes with managers developing specialist services for the communities within which they worked. The structure spread the risks associated with programme management so that local mangers assumed responsibility for their own activities without impinging upon other projects.

The Pittsburgh Manufacturing and Community Development Networks Initiative (PPND, 1996) epitomised the PPND's effort to link community with business interests. Community development corporations in the East Liberty and Homewood–Brushton districts joined with PPND to expand business opportunities so that community and manufacturing networks could involve companies in neighbourhood projects. The manufacturing network extended to include Lawrenceville and the Northside, while the community network concentrated upon broader neighbourhood issues and employment. The networks used community development methods to create new forms of collaboration between companies and local people and they identified common aims and objectives through cooperative actions to address local problems. Manufacturing companies learned the techniques of community development and could apply them to quality improvement and team building. As PPND stated, 'the development of an interactive relationship between the manufacturing and community development networks is an important step in reducing the boundaries between these traditionally separate community sectors' (PPND, 1996, p. 2).

The initiative brought together participants from Carnegie Mellon University, the CDCs, companies and PPND. By 1996, the initiative had enlisted the support of employers, retail firms and banks to work to develop local skills. There was a 'community employment response system' that connected to training agencies and educational institutions where CDCs 'could develop linkages to existing or evolving networks where members share information, resources and access' (PPND, 1996, p. 3). The manufacturing network discussed environmental compliance, company tax, personnel, training, bench marking and accident prevention. These were all interesting to companies, and were issues that involved local people. Companies wanted to support the network because it could help them with real problems. The popularity of the network's 1995 training programme for supervisors brought strong support from member firms, and on-site manufacturing meetings brought the network close to local companies. Such contacts enabled members to exchange information and learn from the experiences of others and companies stood to benefit from improvements in productivity and the creation of a high quality workforce. By the end of 1996, the manufacturing network had an impressive line-up of companies and partners including Nabisco Foods, the Pittsburgh Wool Company, and Best Feeds and Farm Supplies. The Pittsburgh Institute for Economic Transformation, an affiliate of Duquesne University's Graduate School of Business, serviced the network by helping firms to develop strategies to enhance company productivity. The institute built interactive networks of community organisations, manufacturers, and industrial-based firms and it encouraged companies to take a broader view of their interests in neighbourhoods through collaboration and innovation.

By 1998, the PPND therefore supported the development of comprehensive services that addressed urban poverty and encouraged the greater self-sustainability of households through workforce and business development. These activities complemented housing initiatives and the connection of neighbourhoods to the regional economy. During 1997, PPND had an annual budget of $3 million with funding from local and national foundations, banks, and corporations plus income from interest, fees and other activities. Most of the funds went toward support for the CDCs, and the partnership also administered a $2 million Ford Loan fund. Organisational changes in 1997 and 1998 mainly concerned the efficient management of PPND's funding and the support of strong high impact initiatives in the local communities served by the CDCs. Previously, most of PPND's support to the CDCs was through unrestricted annual operating support, but the new approach provided operational support to the CDCs through a Community Economic Development Organization and programme grants. The development organisation provided funds to organisations to engage them in community planning and other activities through the coordination and management of economic development services. The programmatic support provided support to local economic and workforce development and housing. The PPND supported fewer but higher capacity programmes with proficient staff producing high quality services.

The empowerment partnership

The success of the PPND contrasted with the early experience of the Pittsburgh–Allegheny Empowerment Partnership. One explanation for the lacklustre performance of the latter during 1995–6 was that the initiative initially did not fit well with Pittsburgh's existing community culture. The partnership was part of a federal initiative establishing Empowerment Zones (EZs) and Enterprise Communities (ECs). Cities were given the opportunity to compete for federal funds for urban regeneration with EZ or EC status. A Community Empowerment Board, chaired by Vice President Al Gore, brought together federal agencies and in the competitive bidding process local submissions had to show the federal Department of Housing and Urban Development (HUD) that they had innovative strategies that would develop the capacities of communities. The strategies had to provide an integrated approach to economic and community development using a range of public and private funds. The focus was on overcoming economic decline, unemployment and criminality, and as with other initiatives, upon creating public–private partnerships. The Pittsburgh partnership, designated as an Enterprise Community, aimed to expand neighbourhood economic opportunities through sustainable community development and community partnership. The partnership had the support of the City of Pittsburgh, Allegheny County, the cities of Duquesne and McKeesport, and the boroughs of Homestead, Rankin, and West Homestead with the Pittsburgh Department of City Planning acting as the lead agency for the bid. The success of the initiative would depend on the active work of community organisations to involve local people in a variety of economic and community development projects. However, the partnership covered a set of non-contiguous communities with a combined population of 48,713 and their coordination required an ambitious collaborative effort in an area where there was high unemployment and social deprivation. The partners were expected to recognise the interdependence of the communities and also their common interests in achieving the partnership's objectives.

Evaluations by HUD of the EZs and ECs provided good practice guidelines and underlined the factors that contributed to success in urban revitalisation. The 1995–6 performance report for the Pittsburgh EC concentrated on the problems of implementing the original objectives of the programme (HUD, 1997) and referred to the difficulties of creating an effective partnership involving disparate communities where it was difficult to engender a spirit of enterprise and initiative. The HUD evaluation stated that Pittsburgh was one of the poorest performing ECs, and that up to late 1996 the programme had not achieved its objectives in spite of the commitment by Mayor Murphy. The HUD report referred to a general lack of local leadership and recommended that the local partners should make a greater effort to create a leadership focus if the programme was to be effective. The programme had a partnership structure, but it failed to create 'synergy' between very different, often self-contained, communities that had little past practice in mobilising collective resources. These problems were addressed, and a bid in 1998 for Empowerment Zone status came after a period of reassessment and redirection of policy.

Risk and vision

The contingency model predicts that partnerships will reflect the conditions prevailing in particular contexts. The Pittsburgh case shows the changing nature of the political environment that conditioned public–private relationships, especially in the wake of a passing of the old post-war business leadership. Partnerships are varied, and in Pittsburgh business models of management and organisation strongly influence them. This contrasts with Birmingham and Rotterdam where there are strong public sector-influenced structures and processes in community initiatives.

Considering together the partnerships mentioned in this chapter, the City of Pittsburgh pursued a strategic policy based upon anticipating changes in risky regional and local markets, but the strategic direction was channeled through the URA with a variety of partnerships operating locally each with their own strategies and objectives. The role of the URA as a public authority, however, should not be underestimated as a facilitator of major projects and an instrument of urban regeneration and property development. The business leadership tradition meant that the city adopted policies that addressed business interests and the URA strongly promoted development in and around the central business district. The desire of corporate leaders to link communities with the labour requirements of business, and ultimately the regional economy, provided a strong impulse for successful partnerships such as PPND. The sustained growth of the American economy between 1996 and 1998 produced a favourable environment in the city for business and labour, but the vulnerability of the city in the recession of the early 1990s showed how an uncertain market could impede economic growth and affect competitiveness. Business initiatives involving local communities were thus increasingly important, but past recessions inevitably demotivated many poorer communities around the city.

The fragmentation of communities therefore sometimes checked progress, as with the Enterprise Community, where there was reluctance to collaborate and a lack of leadership drive in some neighbourhoods. Partly to overcome the traumas and uncertainties of the market, politicians and business leaders understandably favoured low risk in policy. However, the experience of Los Angeles following the 1992 riots showed the problems that could arise from social conflict and economic adversity. Therefore, the reduction and spreading of risk depend on the restraint of political competition, but the question remains as to whether or not the partnership bodies and agencies of the 1990s have yet developed the most appropriate flexibilities and synergies for promoting competitiveness. Since the 1940s the ACCD played a role in organising and coordinating partners around common objectives, but the link between corporate and political elites was disrupted in the 1970s (Stewman and Tarr, 1982). The disruption of the relationship produced a new and more fragmented corporate elite that took a long time to rekindle a shared vision and common purpose, and this was evident also at the regional level (see chapter four).

6
Birmingham: Strategy and Partnership in the West Midlands

The research in Birmingham was carried out between 1996 and 1998. The first part of the chapter deals with regional policies and networks and shows the political interrelationships between central, regional, and local public agencies and the private sector. The contingent context in Birmingham differs to that in Pittsburgh, and it has produced partnerships that have a different quality reflecting the strong public sector involvement at the regional and local levels. Central government has intervened to maintain order in the economic development and urban regeneration domain and to rationalise partnerships through the establishment of new English Regional Development Agencies. The new political agenda under the Labour government is one of achieving competitiveness increasingly by linking regional and local economic development policies and organisations. This maintains the dynamic of change in regional governance and leads to a stream of policy pronouncements and organisational and strategic changes, but it still accompanies government attempts to order and rationalise several otherwise fragmented policy concerns.

The latter part of the chapter is a case study of the politics of the European structural funds bidding under the former Conservative government. The case illustrates attempts by central government to achieve political restraint in a complex policy setting where interest driven local authorities competed for resources and some opposed central government policies. Central government quite successfully managed to obtain the cooperation of dissident local authorities by effectively managing the main partnership framework and incorporating informal networks. This required centralising fund bidding to reduce local level policy discretion.

Political environment: cities challenged

Birmingham lies at the centre of the West Midlands region (Map 1.2) with a regional economy accounting for 11.8 per cent of national Gross Value Added (Government Office for the West Midlands, 1996, p. 5). At the time of the research, strategy making was the concern of central government, the Government Office for the West Midlands, city, county, and district councils and the private sector. Business plays an increasing role in policy, but even under the former Conservative government public initiatives drove competitiveness policies in the region.

Birmingham lies within a metropolitan area consisting of the local authorities of Dudley, Walsall, Wolverhampton, Sandwell and Solihull (Map 6.1 and Figure 6.1), and the city of Coventry located to the south east of Birmingham. These authorities once made up a West Midlands Metropolitan County, not to be confused with the wider West Midlands region. Today, the old metropolitan county towns and cities compete as independent unitary authorities for central government funds by strongly defending their individual interests. In 1986, the Thatcher government abolished the West Midlands Metropolitan County that up to that time was the top tier strategic authority for Birmingham and the other metro authorities. Thatcher's campaign against Labour party controlled local authorities ostensibly aimed to reduce waste and inefficiency in local government and to make local government more accountable to citizens. However, for Thatcher the metropolitan county represented all that was wrong with Labour controlled city government. Thatcher identified the Greater London Council and the six metropolitan counties, including the West Midlands, with high spending, inefficiency and control by the political left.

Stoker (1991) describes the system established in the former Metropolitan Counties after their abolition in 1986 by referring to the complexity and bureaucratic nature of arrangements that resulted from local government reorganisation. The removal of the top tier strategic counties created problems in important services where responsibilities had been shared between Metropolitan Counties and the lower tier councils such as Birmingham. The new unitary metropolitan authorities, including Birmingham, were given responsibility for services that were subject to various inter-authority agreements and forms of cooperation. Birmingham City Council therefore was a metropolitan authority with unitary powers, but the city coordinated

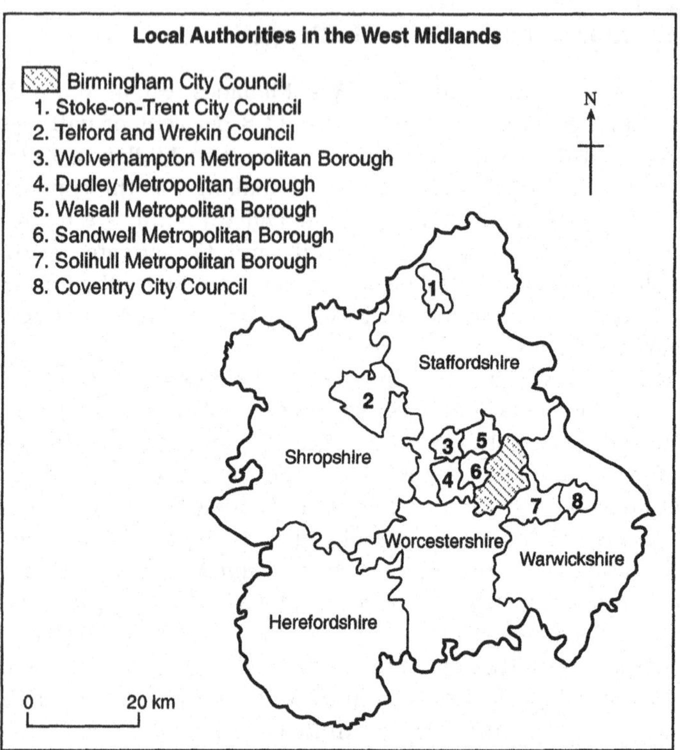

Map 6.1 Birmingham and the West Midlands

Figure 6.1 Population of the West Midlands Region, 1995

County	Population 1995	% change 1981–95
Whole West Midlands Region	5,306,000	2.3
Counties:		
Hereford and Worcester	694,000	2.3
Shropshire	420,000	10.3
Staffordshire	1,056,000	3.7
Warwickshire	499,000	4.8
West Midlands Metro Area:		
Birmingham	1,017,000	–0.3
Coventry	304,000	–5.0
Dudley	312,000	3.9
Sandwell	294,000	–5.2
Walsall	203,000	2.3
Wolverhampton	244,000	–4.8

Source: *Regional Trends* 32 (1997). London: Office for National Statistics © Crown Copyright 1999.

actions with other local authorities where appropriate. However, in local economic development, the absence of an effective strategic authority for the old West Midlands County metropolitan area intensified inter-authority competition for public funds and inward investment and seriously weakened the ability of cities like Birmingham to develop effective long-term regional economic development strategies.

Stoker (1991) shows that Thatcher created a pattern of shared responsibilities through complicated joint-working arrangements and quasi-governmental agencies. Joint committees provided a model for many local services, with variations from area to area, but with administrative arrangements often being removed from the direct accountability of local authorities. The system after 1986 was 'fragmented and confused contrary to government's claims that their reforms would 'streamline' local administration' (Stoker, 1991, pp. 33–4). Local policies strengthened non-elected agencies and created new public–private linkages, and brought an era of 'government by appointment' (Davis, 1996) where central government appointed the chairs of bodies about which the ordinary citizen often knew little.

Coordinating complexity

The complexity of the arrangements that replaced the old metropolitan counties detracted from efforts to coordinate expenditure and provide more efficient urban services. Local governments acted in their own interests at the expense of regional vision and local governments had to administer often uncoordinated programmes and contract out local services to the private sector. In central government, there were also problems in the regions because regional offices of big central government departments operated with rising workloads and poor inter-departmental communications and coordination. The proliferation of economic development and urban regeneration programmes under the Conservative government underscored the need for improved links between the regional offices of central departments and the rationalisation of programmes, such as the European Union structural funds, that channelled substantial resources into the West Midlands.

The Government Office for the West Midlands (GO–WM) was established by the Conservatives in 1994. In all ten of the Government Offices for the Regions (GORs) in England, each with Regional

Directors, the government brought together civil servants from the Department of Trade and Industry, the Department of Employment, the Department of the Environment and the Department of Transport. By 1995, the Regional Director in the West Midlands had responsibilities covering enterprise, employment, environment, transport, industry, training, European programmes and urban regeneration. The Regional Director of the GO–WM reported to the secretaries of state for transport environment, employment education and trade, and he kept close links with other central government departments. Regular meetings of the regional directors and central departments achieved central coordination and accountability to central departments.

Mawson and Spencer (1997) study the relationships between the GORs, central government departments and other agencies by referring to the unusual arrangements making the GORs accountable to central 'parent' secretaries of state. They set regional administrative reform within the context of a move to strengthen territorial management in British government, even though the Conservatives were deeply suspicious of regional government and decentralisation. Even after the establishment of the GORs, according to Mawson and Spencer, line management responsibilities between the GORs and other departments were often blurred. The regional directors brought managers together to coordinate programmes, but the success of such efforts varied according to the programmes concerned. For example, internal restructuring of the GO–WM during 1995–6 reinforced policy integration with the creation of three divisions for housing and regeneration, skills and enterprise, and trade, industrial development, and Europe, and three internal working groups were established to deal with competitiveness policy, urban regeneration, and the environment (GO–WM, 1996, p. 5).

In November 1993, central government announced the introduction of a Single Regeneration Budget which became operational from April 1994 through the GORs. The single budget represented an attempt to overcome some of the problems of fragmented administration in urban regeneration programmes. The budget was part of a package of measures designed to simplify the way government provided support for urban regeneration, economic development and industrial competitiveness and it brought together resources previously administered under separate programmes. The budget included allocations to urban development corporations and to English Partnerships, both of which provided infrastructure investment and prepared sites for development. The Single Regeneration Budget proved to be an effective mechanism for achieving coordination in urban regeneration, and it relied upon competitive bidding for funds between local authorities that enabled the GO–WM to monitor programmes and apply a coherent set of criteria to ensure that monies were spent according to stated objectives. Sponsor ministers worked with the GORs and central government departmental headquarters, and the ministers had responsibility for regeneration matters in their own specified territories.

The GO–WM fostered integrated policies, but regional public–private partnerships were prone to political disagreement and lack of coherence. In the West Midlands, GO–WM officials wanted specialist partnerships to promote competitiveness, regeneration and sustainable development by developing joint plans and local strategies within a broad framework (GO–WM, 1996, p. 4). The officials claimed that partnerships enabled the region successfully to attract substantial investment that might not otherwise be forthcoming, as in the case of the Jaguar automobile plant in Birmingham. The policy reflected central government support for partnerships to improve competitiveness and support the Regional Planning Guidance for the West Midlands that emphasised the inter-relationships between different policies and the importance of public–private sector economic development initiatives. For a Conservative government that was sceptical about formal overarching 'regional strategy,' this represented a commitment that extended the focus of regional policy making by drawing connections between policy initiatives.

Restraining competition and 'Working to Win'

The Working to Win initiative launched in the West Midlands in 1996 represented a concerted attempt by the Conservative government to promote competitiveness and industrial efficiency through ordered public–private partnerships. The problem was that there had been no coherent focus for policy in the regions. The government's approach was, in effect, to restrain political competition through ordered consultation with business and other interests and more effectively address the problems that all regions were facing in remaining competitive within global markets. Several important public–private regional organisations including the West Midlands Development Agency and the West Midlands Regional Forum of Local Authorities had established effective collaborative relationships by the mid 1990s. The West Midlands Regional Economic Consortium had members including local authorities, the Regional Group of Chambers of Commerce, the Regional Trade Union Congress, the Regional Group of Training and Enterprise Councils, the Confederation of British Industry, the West Midlands Development Agency, regional Members of Parliament, Members of the European Parliament and the GO–WM. The partners in the consortium wanted to attract inward investment into the region and maximise regional capabilities in various programmes and 'Working to Win' provided an opportunity.

GO–WM promoted Working to Win following a series of government competitiveness white papers encouraging the public and private sectors jointly to formulate policies and better coordinate central initiatives. However, despite the structural problems of the regional economy, high unemployment, and the loss of manufacturing capacity, the Conservative government refused to proclaim its framework as regional 'strategy' in the West Midlands. Indeed 'regional strategy' remained unpopular with some Conservative politicians who associated it with European Union bureaucracy and socialist-style planning. According to some public officials in the West Midlands, the Conservative attitude made it difficult to focus a competitiveness policy around clearly defined goals even though many of the government's own aims for regional competitiveness had strong strategic implications. Instead, the Conservatives stressed the need for policies to boost economic growth by coordinating individual local initiatives that each had their own objectives but which could work through 'frameworks' such as Working to Win. Such unease with regional strategy was unsurprising even though government ministers did increasingly recognise the value of strategic thinking in planning and urban regeneration. In 1993, the City Pride initiative invited cities to develop 'vision' statements for change. Environment Secretary John Gummer later spoke of the need for cities to adopt a clear direction in line with strategic planning guidelines agreed with the government, and city centres were to adopt strategies for improvement.

New Labour: new regions

After Labour's 1997 general election victory, Prime Minister Tony Blair followed a radical policy of reform programme that was intended to restore the authority of local government and increase policy influence in the regions. The new Labour government's policies combined a continued commitment to public–private partnership and they envisaged extensive local government reform and the development of regional strategies. In 1997, the Labour government proposed a new Scottish Parliament with tax varying powers and a National Assembly for Wales. Blair put the government's proposals to the Scottish and Welsh people in separate referendums and both electorates voted for the proposals. However, the small 'yes' in Wales delayed the campaign for elected regional assemblies for England although the government went ahead with plans for Regional Development Agencies and a new strategic body for London with an elected mayor for the capital. The government also proposed elected mayors for other major cities and an enhanced role for cities in promoting new health, education and economic development initiatives. Central to this approach was the realisation that cities needed stronger political leadership to overcome the weakening of local government that had occurred under the Conservatives.

In the West Midlands, the experience of the GO–WM under the Conservative government had convinced many local government officials and Labour councillors that regional policy and administration should be throughly reassessed. The administrative system was fragmented and often inefficient, and the proliferation of non-elected bodies made administration remote from ordinary people. In economic development and urban regeneration, the allocation and bidding for funds through the GORs was by way of a cumbersome bureaucratic administrative framework involving the government offices and central government departments. The problem after mid-1997, following the election of the Labour government, was to develop a policy that would improve regional level relationships between the GO–WM and local authorities. Local authorities in the West Midlands under Labour control claimed that they had not been involved sufficiently in the early stages of the policy development for Working to Win. With a Labour government in office, there was a strong incentive for Labour councils to contribute more, and Birmingham City Council therefore became a more active participant in regional policy networks and supported Working to Win Two which refocused policy in preparation for the new Regional Development Agency that was later named 'Advantage West Midlands'.

The Labour government advocated greater managerial efficiency, holistic programmes, administrative coordination, stronger local authority leadership and political devolution to the regions. In

December 1997, following extensive consultation with public and private sector organisations, the Blair government published a white paper entitled *Building Partnerships for Prosperity* (Department of the Environment, Transport, and the Regions [DETR], 1997) that detailed arrangements for the establishment of nine Regional Development Agencies (RDAs) in England to become operational during 1999. The Regional Development Agencies would have the same boundaries as the Government Offices for the Regions, except Merseyside, which was destined to come under the North West RDA. Deputy Prime Minister John Prescott committed the government to administrative modernisation and decentralising power away form Whitehall. The government would create a more competitive economy and improve Britain's economic performance through economic development and urban regeneration, and the RDAs would take over responsibilities presently carried out by different national and regional bodies. Some of these responsibilities would transfer to the RDAs from the Government Offices for the Regions and English Partnerships. The RDAs would be non-departmental public bodies, accountable to central government but relying upon business leadership and 'an effective presence from the public sector which is also very active in the regional economy' (DETR, 1997, p. 49).

In 1998, the government published draft strategic guidance for RDAs that aimed to address regional business competitiveness. The RDAs were to have a statutory obligation to formulate regional strategies, and each region would produce strategies that would reflect the particular circumstances facing the newly appointed RDA chairpersons. The strategies would be developed at the regional level, and the guidance was to steer the process within a coherent and integrated framework with a strong emphasis on competitiveness and environmental sustainability. The strategies would provide the basis for economic decision making and partnership in the regions and would link to various central government initiatives in education, employment, and urban regeneration.

Under Labour, the role of business is thus set to expand in regional policy through RDA partnerships, but as under the Conservatives, the government still has to provide the framework for business involvement. The situation in the West Midlands is complicated by different corporate and local government policy strategies that often weaken the regional focus of business, and business involvement in the West Midlands does not match the strong corporate leadership evidenced in Pittsburgh. In Britain, business is to play a major part in the RDAs, and business leaders are being encouraged by the government to assume key roles in education, training and policies against social exclusion. In economic development, business already has a prominent role in the government Private Finance Initiative and in urban regeneration. Business leadership, however, often concerns high level strategy while community organisations frequently are poorly consulted in the regional policy process. Local organisations have less influence in high level regional partnerships despite the activities of organisations such as the National Council for Voluntary Organisations and the Civic Trust. There is therefore a distinction between influential groups that gain access to the policy process regionally and those that have to compete within policy networks to get their demands and interests represented effectively.

Political competition and European Union funds

The following account describes a framework that the West Midlands RDA will take a leading role in substantially modifying. The case, relating to pre RDA days, is included because it provides an example of the strength of hierarchy in the regional partnership where political competition led to low-level conflict that influenced the agenda affecting European Union funding bidding. The case shows how central government and the GO–WM strongly influenced regional partners through the competitive bidding regime for European Union structural funding. The research was conducted between 1994 and late 1996 during the lifetime of the former Conservative government, and it concentrates on funding allocations for specific local projects with reference to setting broad regional strategy for European Union programmes. The case shows that political tensions between the Labour party-run Birmingham city council and the Conservative central government arose under a highly competitive fund bidding regime.

European regional funding under the Conservative government was a stage of intense political competition between central government and local authorities. The British government eagerly accepted European Union regional funding even though government ministers had called for an assessment of the effectiveness of European programmes and their administration (Audit Commission, 1991). The European Commission supported economic restructuring to overcome unemployment and financially assisted regions to promote regional 'cohesion', but the Skeptical British criticised European regional aid for bolstering costly projects that provided only low economic returns. Until early 1995, Bruce Millan was the Commissioner responsible for running the regional policy directorate of the European Commission, known as DG16. He was a former British Labour party government minister who advocated a strong regional role for the European Union spearheaded by DG16. His policy targeted regional economies especially through the European Regional Development Fund (ERDF), and the European Social Fund (ESF). The ERDF provided assistance to regions for economic development and social cohesion, committing resources to new infrastructure, community development, job creation and business support. The ESF dealt with employment issues and assistance for vocational training and other job enhancing initiatives.

Bruce Millan's successor, Monika Wulf-Mathies, recognised that the effective use of these funds depended upon clear objectives, decentralised management, effective strategic public–private partnerships and good programme monitoring and evaluation. But a major cause of political competition and conflict between regions in the European Union concerned the designation of 'objective status'. Since the structural funds favoured the less developed regions of the European Union, they had theoretically to produce maximum economic benefits for the regions. The objectives aimed to achieve this through the geographical and functional targeting of financial resources for economic and social development. Objective 1 resources went to the less developed regions while Objective 2 status was conferred on areas that required industrial conversion following the decline of employment. Other objectives covered rural programmes and non-geographically targeted programmes for unemployment, training and agricultural restructuring.

Seeking political restraint

Benington (1994) argues that domestic programmes in Britain adapted to European Union regulations and procedures and that British local authorities became more active in European networks. Complex interagency relationships brought new involvements for public sector organisations working with private sector interests in influential regional and local partnerships (Stewart and Stoker, 1995; Leach, Davis, and Associates, 1996; Jacobs, 1996). Partnerships developed strategies and implemented policies that were not always clearly defined, but partnerships were at the heart of European Union economic development initiatives (Andersen and Eliassen, 1993). The European Commission's evaluation of the structural funds for 1989–93 therefore confirmed the enormous diversity of programmes that depended upon public–private sector cooperation (European Commission, 1995b). Middlemas (1995) highlights the role of informal policy making in the European Union where networks played a central role in the development of policies favouring business and commercial innovations. He argues that European Union policy mobilised diverse corporate and local authority interests that lobbied the European Commission. Similarly, Leonardi's (1995) interpretation of regional economic development concentrates on policy initiatives using new types of delivery mechanisms through vertical, horizontal, intergovernmental, inter-sectoral, inter-functional and interpersonal networks. The allocation of European Union funds in the West Midlands was a problem for central government precisely because of the existence of such complex organisational relationships that increased the chances of political conflict and inter-organisational rivalry. This was evident in Birmingham in the mid-1990s where European Union programmes operated at four levels with corporate and local government demands articulated at each level. The top level involved national governments working with, and trying to influence, the European Commission. At the national level, the British government was responsible for European programmes, negotiating with local authorities and developing broad policy objectives. At the regional level, the GO–WM allocated funds in consultation with various groups, and at the base level local authorities linked with local community organisations and others to bid for funds and set policy priorities.

The geographical areas designated for Objective 2 funding under this system in the West Midlands covered a population of 3.04 million including Birmingham. The GO–WM facilitated the regional and local level partnership framework for the Objective 2 areas, and for 1994–6, a commitment of 938.4 million European Currency Units (ECUs) of public and private funding went to the Objective 2 programme in the region including the ERDF and ESF. However, the formulation of European regional funding priorities in the West Midlands frequently evaded close public scrutiny and government departments responsible for European regional policy developed remote, bureaucratic and complex structures. This remoteness was especially manifest as the Conservative government sought to rationalise regional administration and exert control over regional economic policies and funding. The government tried to restrain political competition through top–down intervention in regional structures that were strategically important for the GO–WM in implementing successful European initiatives. Coherence was important given the Regional Economic Consortium's (see above) attempts to attract inward investment into the region and to maximise the region's capabilities in European programmes. Different initiatives operated through the consortium's Regional European Strategy and regional Innovation Strategy and by way of the competitiveness agendas of organisations and partnerships such as the West Midlands Developent Agency (1995) and the West Midlands Regional Forum of Local Authorities (GO–WM, 1996, p. 13).

Efficiency as a theme encouraging restraint

Efficiency was also an important theme in the policy discourse. Different partnerships encouraged innovation, brought businesses together and enabled public authorities to devise common objectives. The European Commission strongly emphasised regional economic cohesion, the reduction of disparities between European regions, and the development of trans-European transportation, telecommunications and energy infrastructures, and this involved many interests. The British government had little enthusiasm for regional cohesion and, like other European governments, adopted a cautious position on the development of trans-European infrastructures. In the administration of the structural funds however, efficiency and value for money criteria were crucial, and the British attempted to avoid the problems that had beset other governments in trying properly to control and monitor European Union funds. The GO–WM closely supervised European programme regimes in the West Midlands, and the British government sought to rationalise policy by attuning European Union regional initiatives and operational guidelines so that they were in line with those in domestic urban programmes. The Conservative government therefore supported changes in programmes to favour administrative rationalisation and tight financial control within the European structural funds.

Using European Union guidelines, the British government supported the drafting of a strategic regional Single Programming Document for European Union initiatives in the West Midlands. It replaced a framework that many politicians and public officials claimed afforded a bigger role to individual local authorities in the setting of specific funding priorities. The old framework relied on the subregional allocation of funds whereas the new Single Programming Document placed greater emphasis upon region-wide priorities for the West Midlands as a whole. Labour controlled councils, including Birmingham, were critical of the Single Programming Document for weakening their specialist local contribution to the allocation of structural fund resources. Labour critics argued that the British government had devised new fund bidding arrangements and programme delivery regulations for the ERDF and ESF to make local authorities respect strong central supervision and that local authorities and their partners had to meet tight deadlines and stringent conditions when submitting competitive funding bids on a region-wide basics.

Mayes (1995) argues that the British government sought to achieve efficiency in European Union programmes, but without allowing regional autonomy. Public–private partnerships were useful for GO–WM officials because they were often adaptive and unconstrained by formal organisational traditions and rules, but this meant that they could also change to fit in with GO–WM expectations. Policy therefore was much less about partnerships contributing to a regional vision of a new Europe of the regions than about their contribution to efficient and effective working. Local government permanent officials also liked partnerships because they employed the talents of expert council officials and extended the roles of local councils in economic development, so this provided a common interest in joint working between local authorities and the GO–WM.

Defending interests

How did regional policy changes affect Birmingham specifically? Birmingham City Council's European and International Affairs Task Force was as committed as the Conservative government to efficient and effective programmes. The task force was responsible for all European funding initiatives, the monitoring and evaluation of programmes, and the provision of policy advice to the city council. It participated in the Eurocities network to influence European Commission policy on the structural funds, and through such networks the task force articulated Birmingham's reservations about the new regional Single Programming Document. Birmingham City Council European regional strategy in the early 1990s had placed emphasis upon the needs of specific targeted areas with a special focus on Birmingham and the adjoining Solihull Metropolitan District. The policy was contained in a 1994 operational plan that developed a subregional perspective for Birmingham within the wider

West Midlands. However, as the then central government Department of the Environment (DoE which later formed part of DETR) regarded this subregional approach as a less viable than the development of broad priorities for the West Midlands as a whole, there was bound to be disagreement. Birmingham's subregional approach failed to get the approval of the DoE, but city officials continued to argue that European funding should be spent in Birmingham and Solihull on the basis of a subregional vision of partnership for a vibrant growing economy. Birmingham's approach drew upon the work of a multi disciplinary team drawn from the main partners within the Birmingham–Solihull subregion. The approach provided a framework for fund applications under Objective 2 of the structural funds that had been agreed after discussions between Birmingham City Council and Solihull Metropolitan Borough Council, Birmingham Heartlands Urban Development Corporation, and representatives of local educational institutions and the Birmingham Training and Enterprise Council. These partners were committed to joint actions by the local authorities, national and local governments, and other agencies and organisations.

During 1994, disagreements between the DoE and Birmingham City Council intensified. The DoE wanted its regional priorities in a form that the city council claimed would undermine an integrated approach to allocating European funds. The DoE supported area bids from within the whole West Midlands with cities competing for funding, but Birmingham preferred clear detailed subregional selection criteria for projects, but allowing for the setting of targets within the wider West Midlands strategy. The DoE underlined the need for region-wide priorities for business support, training, and economic development, but the DoE approach, according to city council officials, reflected the territorial demarcations of government departments ill-disposed to a framework that favoured local authorities. Nevertheless, by the end of 1994, the DoE view prevailed, despite attempts by Birmingham to sustain what officials called a 'shadow' subregional network. European Union initiatives therefore now operated within a framework of formal consultation around the Single Programming Document priorities for Objective 2. The Single Programming Document provided the formal partnership vision for the Objective 2 area and set out the rationale for programme policy themes and priorities (GO–WM, 1994). In addition, English Partnerships and the Birmingham Heartlands Development Corporation were important in establishing local partnerships for the development of major projects using European funds.

The priorities in the West Midlands Single Programming Document covered the development of the region's productive activities, transportation and communications, business development, image improvement, tourism and research, training and development. For 1994–6, a commitment of 938.4 million ECU of public and private funding went to the Objective 2 programme including ERDF and ESF. In the West Midlands, the bidding guidelines and evaluation reinforced top–down policy control especially as the GO–WM encouraged administrative rationalisation and local authority adaptation to programme funding regimes. The important processes were associated with, but aside, mainstream local government, and this situation led to demands for improved public accountability. In Birmingham, the system therefore involved many actors operating within the ground-rules heavily prescribed by central government departments and the GO–WM, and local authority officials had to respond by taking decisions on technical issues and fund bidding. By late 1996, the process worked through inter-linked and specialised mechanisms that required the GO–WM to order the policy process by encouraging rationalisation, programme efficiency and local compliance with funding regimes. It also strengthened what most Labour councillors regarded as the bias in favour of the policy priorities of central government and the inequities of competitive bidding.

In 1996, the allocation of ERDF funds in the West Midlands therefore depended on rationalised but upwardly responsible series of local partners, the Single Programming Document, and a monitoring committee with oversight over the allocation of project funding and programme implementation. Local partnership groups operated within five subregions, each responsible for local policy, the identification of local needs and the development of local partnerships. The formal regional partnership involved the GO–WM, local authorities, Training and Enterprise Councils, colleges, companies and voluntary community organisations. A programme secretariat was part of the GO–WM with authority to appraise project proposals from the local partnership groups. The final decisions on projects came after projects went before the Department of Trade and Industry, the DoE and the Heritage Department.

Political competition and control

European Union programmes in the West Midlands contained a strong element of top–down influence despite the fuzzy networking between different groups of organisations bidding for funds around the region. Some public officials felt some of the more loosely networked arrangements around European Union funding, such as the 'shadow' network, consolidated and adapted to the 'official' framework as partners established formal links with the GO–WM. Partners tended to respond to top–down interventions especially as they were influenced by GO–WM programme guidelines. The European programme process therefore developed more centralised coordination as important partners established closer relationships with the GO–WM that controlled bidding and monitored operational programme management. However, partners did increasingly influence the 1997–9 Single Programming Document through various working groups.

Partners had different levels of commitment to networks (Godfroij, 1995) and they could not be expected always to pull in one direction. Local authorities sometimes challenged centralised restraint to such an extent that some public officials claimed that the formal GO–WM framework did not constitute a proper partnership. Many Labour councillors regarded European bidding as influenced by central government policies with objectives, competitive criteria and project delivery methods looking remarkably similar to those in domestic government programmes (Birmingham City Council, 1994). Therefore, local authority officers sometimes tried to make the process more political by localising decision making by maximising local authority involvement and through informal lobbying in wider regional networks that could influence GO–WM.

Regional governance

Le Gales and Lequesne (1998) stress the complexities and contradictions of change by regarding institutional reform and complex governance as part of the transformation of modern capitalism. Le Gales and Lequesne (1998) assess the factors that produce regional change arguing that during the 1970s regional institutions were essential for the regulation of economic activity and the effective implementation of welfare state policies. However, it was difficult to strengthen regional institutions because resources were limited and there was political opposition to big government. The experiences of the years of Conservative government in the West Midlands bore this out as regional political interests were poorly organised during the 1980s and regional institutions and local governments did not keep up with economic regionalism and the transformation of regional production and trading patterns. Local authorities, including Birmingham, took part in complex regional–local networks

(Leach *et al.*, 1996) to ensure access to centrally distributed funds, and this produced distinctive clusters of organisations cutting across different policy domains such as training, housing, health and social provision. This strengthened the role of public organisations as facilitators for the private sector, policy coordinators and providers of public funds, but it also meant that regional networks were politically competitive and prone to conflicts with the centre.

7
Birmingham: Partnership and Community

This chapter covers local economic development and urban regeneration in Birmingham from research conducted between early 1996 and September 1998. The focus is on citywide strategies and public–private partnerships that involve local community groups. It shows the constant revision and development of strategy in a city with many interrelated partnerships and policy initiatives. The attempts to coordinate initiatives have continued under both Conservative and Labour governments giving an impression of endless policy change, strategic reassessment and programme reorganisation.

Community initiatives are important in Birmingham where black, Asian and other ethnic minorities make up 21.5 per cent of the population, and where some neighbourhoods rank among the most deprived in Britain. The issues of social exclusion and the need to improve the economic prospects of citizens directly reflect the concerns of policy makers in the European Union. The Birmingham and Rotterdam approaches thus relate to the emerging urban agenda in Europe (European Commission, 1998) through the need to develop local programmes that are sustainable and empowering and which connect local people to regional economic growth.

The chapter starts with the recently terminated Birmingham Heartlands Development Corporation where restrained political competition underpinned an economic development strategy based on business investment and targeted growth. The development corporation model, conceived under the Thatcher government, created unelected local boards using central government funding in specially designated areas that, for economic development purposes, were outside local government. The chapter also covers Birmingham City Council's Economic Development Strategy which under the Labour government assumed a broader regional orientation and supported holistic programmes and quality-driven 'Best Value' local services. The chapter details an urban regeneration partnership funded from the Single Regeneration Budget under a central government-defined framework that allowed a degree of local initiative and policy discretion. The chapter thus illustrates the complexity and variation in local economic development and urban regeneration, and it traces attempts by the city council to work more strategically. It indicates how citywide strategy is but one context for partnership working rather than a rigid template.

Political restraint

The Birmingham Heartlands, to the east of the central business district, long provided the city with a symbol of the depth of the decline of the West Midlands industrial base. The Heartlands area once was a mainstay of the city's industrial capacity, but by the 1980s it represented declining manufacturing industries and dramatic economic dislocation. Like the industrial Black Country north west of the city, the Heartlands had old smokestack factories, power generation sites and poor housing that by the 1980s formed a core of physical dereliction. At the end of the 1980s, Birmingham City Council and the private sector identified the Heartlands as an area in need of urgent attention. In 1992, the Birmingham Heartlands Development Corporation (BHDC) was established to build upon existing initiatives by developing 1000 hectares of land within a jurisdiction comprising of manufacturing and derelict land sites. The area included some of the largest of the industrial sites in the West Midlands, once occupied by manufacturers famous in British and overseas markets.

The Heartlands development corporation contributed to infrastructure development, job creation and training. It was a short-life corporation with a remit to take it to 1998. Margaret Thatcher's Conservative government created the first urban development corporations in the early 1980s as a challenge to Labour Party controlled local authorities that she accused of lacking initiative. Central government favoured conditions conducive to economic growth and the emergence of an enterprise culture. Urban development corporations were to cut bureaucracy and quicken urban regeneration by concentrating responsibility for economic development with government appointed local boards using public funds to 'pump prime' projects.

Not surprisingly, Labour Party councillors nationwide were initially reluctant to support such initiatives because of the supposed threat the urban development corporations posed to local authority economic development departments. Labour councils criticised the corporations for their bureaucratic organisation and their poor representation of local interests, and Labour associated them with Thatcher's drive to reduce the powers of local authorities in favour of business. However, attitudes gradually changed as the tangible benefits of the corporations became evident and as Labour councils moderated their politics. The corporations made substantial improvements to run-down urban areas by attracting private sector investment and creating jobs, and this benefited big Labour Party controlled cities such as Liverpool and Manchester. By the early 1990s, even left wing Labour councillors recognised the advantages of having a local urban development corporation that could attract central government funding for prestigious projects. Labour politicians became more favourably disposed toward the corporations and sought to influence them by maximising councillor representation.

In Birmingham, the development corporation established a close relationship with the city council and it avoided the political conflicts that troubled the earlier established corporations in Liverpool and London. The council learned from the past, and the city had good representation on the Heartlands board through its involvement with previous initiatives in the area. The close involvement of the city council helped to overcome lingering suspicions in the Birmingham Labour Party that the corporation would threaten the city as a leading planning authority. By the early 1990s, local politicians accepted that political confrontations with central government were counterproductive, and they perceived the advantages of working with the private sector during an economic recession. The Heartlands board consisted of 12 members in 1997, five of whom were city councillors, with others from the private sector, nonprofit organisations, a local community regeneration programme and the National Health Service. The members of the board had overlapping interests and experiences in a range of nonprofit and for-profit private sector organisations. The board's composition suited the city council because it gave the city policy influence and allowed for community consultation. The development corporation therefore consulted with a range of interests and employed professionals familiar with competitive markets and the needs of national and local companies. The management structure of the corporation also ensured that the city council's Director of Planning acted as advisor to the board and that the chairperson of the corporation's planning board sat on the city's planning committee.

As with urban development corporations in other British cities, the Heartlands corporation acquired land for development, initiated land reclamation and building construction, and provided land management and other services. The corporation's aim was to assist the restructuring of the industrial base of the Heartlands by way of land clearance, revitalisation, grant assistance and encouraging cooperation and coordination with the city council and other local public agencies. The objective was to work with the private sector to make things happen by indirect actions and making large direct interventions when market conditions made them necessary. Officials therefore boasted that the corporation had secured the base of large firms through funding for Jaguar automobiles, and that SP Tires and Dunlop Aircraft Tires had their factories rebuilt with large grants. Corporation officials claimed that

> LDV, the largest British owned vehicle manufacturer has been supported in its rise from the ashes of the DAF receivership, with help from BHDC, the Department of Trade and Industry, and the city council. GEC Alsthom is building new trains for the new railway companies. A host of small companies have been helped, industrial land reclaimed, and new industrial development encouraged (BHDC, 1997, pp. 1–2).

For the financial year 1996–7, the BHDC received £8 million from central government, accounting for 46.3 per cent of the corporation's income. The rest of its income came from the European Regional Development Fund and capital receipts.

The variety of projects supported indicated the links between local community regeneration and local authority economic development policies, but the passing of the corporation marked the end of an era in urban regeneration. It represented the final passing of Thatcher's urban development corporation model at a time when Labour's holistic policy approach entered the policy agenda, and as city council and English Partnerships continued with development initiatives in the Heartlands, envisaging a strong regional role for the area in the city's future. The urban development corporation model had provided rapid development solutions, but it also produced a fragmented pattern of development that reflected a responsive approach to changing market conditions. Thatcher's model relied upon restrained political competition that obliged the city to work within parameters defined by central government. It was a model under which the city accepted a development process in the Heartlands that, despite city representation, was aside from the city's own longer term economic development strategy. The model was of significance economically, but the engagement of local people was limited and localised compared to the grander regional community economic productiveness and empowerment strategy emerging in late 1990s European Union and Regional Development Agency policies. The physical pattern of economic development in the Heartlands reflected the market-responsive philosophy of the corporation that produced a landscape of architecturally diverse and uncoordinated developments of different functions and scales. Political restraint in the Heartlands thus allowed a form of urbanism built on policies that valued short-term investment opportunities, quick development fixes and social interventions at the margin.

Policy coordination

Well before the election defeat of the Conservative government in May 1997, the ruling Labour party on Birmingham City Council sought better coordinated local economic development programmes. The council's broad economic development strategy focused on the construction of prestige projects in the central business district and the regeneration of deprived communities. The city Economic Development Committee supported policies protecting local industry, stimulating economic growth and encouraging inward investment. It was a pragmatic approach, driven by the political desire of the Labour leadership to enhance Birmingham by offsetting what they regarded as the adverse consequences of Conservative policies. The strategy concentrated on issues directly affecting Birmingham which sometimes detracted from a detailed assessment of wider regional considerations. However, with a Labour government in office, the city concentrated more on a coordinated strategy for the West Midlands. The prospects of proactive regional work, and a change in emphasis by the Government Office for the West Midlands (GO–WM) toward acceptance of an explicit regional strategy (see chapter six), encouraged the city to develop the wider regional perspective in anticipation of the full operation of the West Midlands Regional Development Agency in 1999.

In locally targeted programmes under the Conservatives, central government initiatives in training and education, business support and enterprise increasingly intermeshed within local urban regeneration partnerships. Public sector intervention, especially when it stimulated private initiative, provided a point of reference in the local political relationships between agencies and groups. Much of the discourse in the politically restrained networks that formed was about managerial efficiency, the achievement of outputs and partnership synergy. However, Jones and Ward (1997) revealed the problems facing urban regeneration organisations under the Conservative government and how partnerships failed effectively to respond to changing economic conditions and competitive pressures. Jones and Ward argued that the Training and Enterprise Councils (TECs) and the Single Regeneration Budget relied on public and private sector support but did not adequately cater for local needs. Management problems were arising from the fragmentation of initiatives, the lack of strategic coordination, and problematic relationships between different local government departments. In particular, Jones and Ward argued that the TECs, and the government Business Link initiative, lacked public accountability and displayed a strong policy bias that neglected community interests. However, the TECs, and a plethora of quasi-governmental bodies, also failed to satisfy the needs of the business community because of their managerial complexity and lack of flexibility. Jones (1995) referred to the lack of accountability of the TECs, their exclusive top managerial selection processes, poor complaints procedures, and difficulties in obtaining the trust of their partners. Jones and Ward viewed the TEC National Council's own recognition of the difficulties, and their findings supported other research revealing local services facing severe problems as they fragmented and privatised (see also Chisholm, 1997). It was especially important to note the managerial problems of the TECs, Business Link, and the Single Regeneration Budget because of the interrelationships between them. In Birmingham, the TEC played an important role in partnerships in education, training, and business support. Business Link involved the Birmingham Chamber of Commerce with the TEC and some funding from the Single Regeneration Budget for business support services. The partnership process therefore brought together local agencies that had strong interests in rationalising management and organisation through coordination, resource sharing and networked service delivery mechanisms. However, demarcations led these bodies to compete for funds and defend their own operational interests. Under the Conservative government, coordinated programmes were therefore important for the city council and others to assure Ministers that they could more effectively implement local programmes.

City officials wanted the efficient and effective delivery of policies that were internally consistent and appropriate for dealing with changing external conditions. They sought the coordination of policies through good working relationships with the organisations representing business and local communities. The *Economic Development Strategy for Birmingham 1996–99* contained strategic policy objectives taking account of the Economic Development Department's annual assessment of the city's economic record. The annual reviews provided three-year projections of trends that influenced the objectives for economic development and other city policies. The review process monitored changes affecting Birmingham in the national, regional, and international economies and the policy impacts on local communities and businesses. In this way, the strategy focused on local people and the need to foster partnerships between the public, private and nonprofit sectors. It covered practical issues including the need for an effective property development strategy recognising the shortage of prime development sites in and around the city.

City Pride

The City Pride initiative provided a forum for the elaboration of local economic development and urban regeneration policy involving the city in extensive consultation with business, community and public organisations. City Pride originated from a Conservative central government initiative in November 1993 designed to improve the delivery of urban regeneration programmes through better coordination between agencies. The Conservative government invited Birmingham, Manchester and London to define ten-year visions for urban regeneration with local authorities taking lead roles in coordinating policies. Clarke and Prior (1998, p. 4) argue that City Pride produced a 'more corporate conception of urban governance and policy' that marked a shift in Conservative government thinking away from an obsession with market forces. The Conservatives encouraged cities to frame policies that would more effectively focus objectives and better manage urban regeneration through a multiplicity of related programmes. Birmingham extended community consultation to develop a new strategic vision, and Clarke and Prior (1998) describe the commitment of the city to an important interagency management group that steered City Pride. The group included City 2000, a business interest group and the Birmingham Chamber of Commerce.

City Pride supported programme coordination for the Single Regeneration Budget and other initiatives to establish better policy linkages and programme performance. City Pride, as it developed, enabled Birmingham to develop a coherent perspective to overcome fragmentation, although it introduced a new level of policy and organisational commitment in city policy. The initiative provided a long-term context for policy development with a City Pride Board supporting the initiative. The Board had members representing the Chamber of Commerce, the Health Authority, the Police, Birmingham University, City 2000, the TEC, City Challenge, trade unions, tenants, youth, business and voluntary groups, and city councillors. The board linked to public–private partnerships, European programmes, central government and the city Economic Development Department. The first City Pride prospectus appeared in 1994 with a refined development strategy. A second prospectus, in April 1995, represented the outcome of consultations involving all stakeholders. It covered the local economy, Birmingham as a regional capital, social conditions, youth policy and community regeneration. Figure 7.1 shows the relationship between the city's economic development strategy and City Pride as it was presented in the 1996–7 economic development strategy.

Figure 7.1 shows how the economic development strategy related to City Pride, although the framework suggests that the city designed the approach to fit central government expectations about how the initiative should work. Nevertheless, the economic development strategy at that time integrated five City Pride core objectives and set out the ways of carrying out policies in a coordinated manner. Interrelationships between programmes in economic development and urban regeneration were important, including the European Union structural programmes, the Single Regeneration Budget, the old City Challenge programme, Housing Action Trusts and the Heartlands development corporation. Figure 7.1 shows city council commitments set within a framework that envisaged cross-linkages between economic development, City Pride and council departments. By early 1997, the city therefore had an emerging strategy based on the recognition of these links. The Economic Development Department liaised with the City Pride Board and the Policy Division

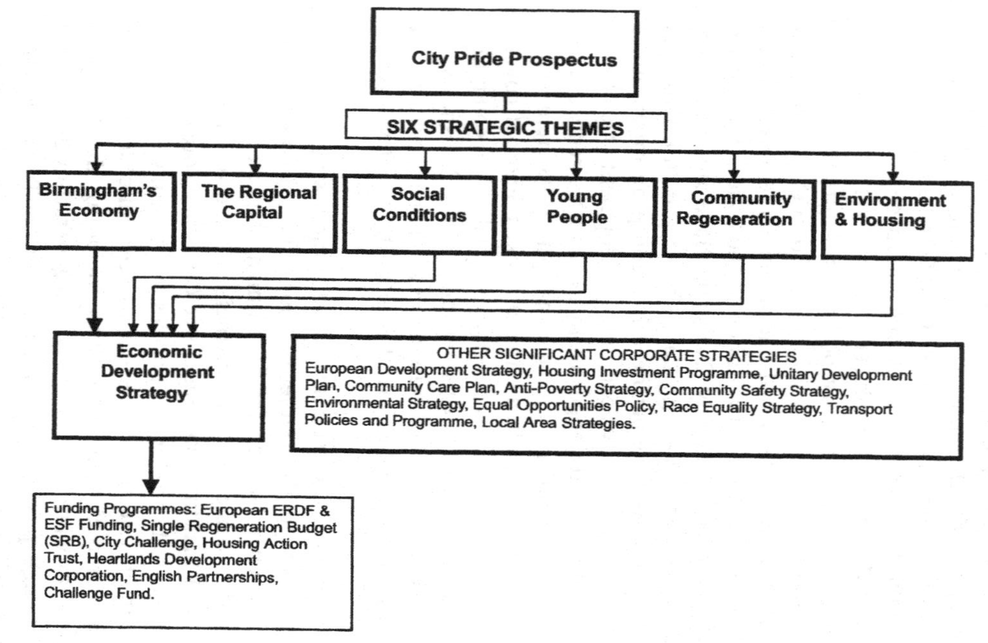

Figure 7.1 Birmingham Economic Development Strategy, 1996–9

Source: Adapted from Birmingham City Council Economic Development Program 1996–7.

of the council that took direct responsibility for the monitoring of the Single Regeneration Budget programmes. Later, City Pride featured less prominently as an integrating focus in Birmingham. City economic development policies developed from the earlier foundations with City Pride becoming a more targeted strategic initiative concentrating on specific projects. The Labour government's agenda featured new initiatives that demanded a revision of the way strategic integration could be achieved.

Towards holism

One aspect of strategic refocusing related to regional policy. Up to 1997, Birmingham City Council was active in regional networks and lobbied hard for European Union funding, but the city emphasised its own economic development strategy and major projects of regional significance. The Labour council remained sceptical about the Conservative government's attitude toward Birmingham while wanting to satisfy the government that the city could effectively manage centrally funded programmes. With the election of the Labour government in 1997, the blurring between regional policies and local economic development strategies through the West Midlands Regional Development Agency (see chapter six) and the collaboration of local and regional organisations provided the basis for new holistic approaches. The Labour government initiated important new programmes nationally including Health Action Zones and Education Action Zones to 'join-up' policies and different agencies in social welfare, education and related areas. Programmes would join-up initiatives to combat youth unemployment, improve housing, raise education standards and improve health and welfare provision (Department of Health, 1998). The government also supported stronger local government leadership, with the prospect of elected mayors for major cities and the internal reorganisation of local authorities to support their enhanced leadership roles. The Labour government's Social Exclusion Unit underlined the need for holistic solutions and innovative policies to overcome community problems. For the Social Exclusion Unit (1998), urban regeneration had to be comprehensive, with long-term interventions to empower communities. In 1998, the government announced 'Pathfinder' local authorities, including Birmingham, to develop innovative projects through a New Deal for Communities programme and initiatives in local services.

Birmingham city officials and the Economic Development Committee recognised that their local strategy depended upon an assessment of how local organisations could be involved in these ambitious policies. Many organisations with interests in urban regeneration still worked with little coordination and without developed shared policy visions. In July 1997 under a New Labour government, Birmingham City Council met with eighty delegates from the public, private and voluntary sectors to shape priorities. The parties agreed a 'shared understanding of the forces at work in the local economy' (Birmingham City Council, 1997, p. 2) and to appraise the existing economic strategy. The conference held workshops to discuss issues that affected future economic development in Birmingham with a particular emphasis on the need to overcome social exclusion. Delegates discussed a draft Economic Strategy Action Plan formulated in May 1997 and suggested a range of objectives to support the emerging shared vision. The Labour government's commitment to combat social exclusion gave the city an opportunity to expand the vision with new crosscutting policy themes reflecting government thinking on innovative local economic development and urban regeneration.

The city Economic Development Strategy for 1998–2000 provided a stronger regional theme following consultation with the chamber of commerce, the TEC and other stakeholders. There was a clear vision of Birmingham (Birmingham City Council, 1998a) as a world class city and a commitment to strengthening regional and local partnerships through an action plan connecting initiatives to local communities. The strategy involved policies for training, business competitiveness and investment in sustainable development and infrastructure. There was a recognition of the complexities of the problems facing the city and the need for comprehensive policies to address multi-faceted problems and community needs. The Action Plan (Birmingham City Council, 1998b) extended the vision by committing the council to working with various partnerships to regenerate the local economy. One of the major themes in the Action Plan was the development of the city's information technology and communications infrastructures and the development of new services and job opportunities in high technology ventures. This required the establishment of 'a local partnership of information and learning providers' to promote the development of high-tech infrastructures and networks (Birmingham City Council, 1998b, p. 34).

From a separate consultation with community organisations,

originating before the Blair government, the city produced a Community Economic Development Framework that specifically aimed to overcome social exclusion and community disadvantage. The framework was strategic and flexible, and connected with the government's New Deal policy to get young unemployed people into work. In addition, a 'Local Involvement, Local Action' (LILA) initiative provided a broadly based community networking and consultative framework connecting the city to local organisations.

Complex partnership: the case of the Challenge Fund

In line with attempts to achieve a more holistic approach after 1997, the city assessed the roles of existing programmes that involved local groups in collaborative initiatives. The following case details some of the important lessons of partnership working at the local level.

The Birmingham Single Regeneration Budget Challenge Fund (SRBCF) had demonstrated the complex interrelationships between public and private organisations working in diverse programmes. Existing SRBCF projects had to be accommodated under New Labour, although they originated under the Conservative government. Birmingham, like other cities, inherited a system from the Conservatives where SRBCF funds were allocated through competitive bidding. When a bid was successful, a community board, company, or other appropriate partnership management group took responsibility for the local programme.

The Conservatives, by channelling urban regeneration money through the Government Offices for the English Regions, used the SRBCF to rationalise funding. The SRBCF therefore distributed funds for urban regeneration following competitive bidding for money between English local authorities and other organisations. Five 'rounds' of bidding took place between 1994 and 1998, with £3 billion allocated in England in the first three rounds with an anticipated £5 billion of private sector investment (DETR, 1997, p. 21). The Labour government maintained the SRBCF, but sought a better fit between local programmes and national policies on competitiveness, social exclusion and welfare to work. In 1998, the Labour government granted two new SRBCF programmes to Birmingham covering community safety and measures to overcome social exclusion. The SRBCF depended on GO–WM and city council scrutiny. However, in spite of the GO–WM control over the distribution of core funding for the SRBCF, there was scope for local policy discretion. The programme allowed local authorities to set objectives and develop projects in line with their own priorities provided they did not diverge from certain programme guidelines. The GO–WM and central government bidding framework therefore allowed some variety in local programmes, so once funding became available local partnership managers and local authorities could develop locally relevant projects.

The account that follows deals with Birmingham City Council's 'round one' SRBCF programme. The successful programme bid received initial funding at the end of 1994 worth £28.3 million. The bid divided three ways between an Enterprise Link initiative, an Education Business Links scheme and an 'area' initiative in the Saltley and Small Heath districts of the city. The Enterprise Link provided a citywide thematic element in the overall programme prepared by Birmingham TEC, the city council, Birmingham Chamber of Commerce and other organisations. The Enterprise Link was to operate through a contracting organisation known as the Economic Development Partnership joining the city, the chamber of commerce and the TEC. Prior to SRBCF funding, the TEC and the city council had supported a complex network of separate service level agreements and contracts involving Birmingham Enterprise Network, public agencies and various nonprofit organisations. This provided an 'enterprise infrastructure' that included the TEC's Business Start Up programme that had received government urban regeneration funds since 1994. The Enterprise network coordinated local provision for new enterprises, business advice, property availability and funding information. Enterprise Link could thus network with agencies to support existing local enterprises. It also served the special needs of the self-employed and ethnic minorities through 'provider networks' promoting marketing activities, specialist enterprise services and consultation.

The Saltley–Small Heath Regeneration Partnership to 1997

The elements of the overall programme contributed to economic competitiveness by raising awareness in communities of business opportunities through a 'client-responsive' enterprise service. The local 'area' initiative was a regeneration partnership in Saltley–Small Heath with the city leading the funding bid with the support of partners including the neighbourhood based East Birmingham Community Forum, education institutions, local housing associations and local companies. The Saltley–Small Heath Partnership covered two city council renewal areas declared before the establishment of the programme, and the Small Heath ward. The Saltley–Small Heath Partnership served a population of 50,000 in an area adjoining the boundary of the Birmingham Heartlands Development Corporation, a City Challenge programme and a Housing Action Trust. These together formed a large zone of special urban regeneration initiatives within the city. Each of these regeneration initiatives worked as self-contained programmes, although there were both informal and formal links between them.

The Saltley–Small Heath Regeneration Partnership ran the local area programme that received £12.8 million funding under the SRBCF. The background policy strategy for the Saltley–Small Heath Partnership appeared in the original funding bid document that described the area in south Birmingham 'with a population and economic base the size of a small town' (Birmingham City Council, 1994). Saltley and Small Heath had a young population with 34 per cent aged 15 and under. Social problems seriously disadvantaged this community where so many lived in poor housing and suffered from serious economic disadvantage.

In local SRBCF programmes, including Saltley–Small Heath, a 'delivery mechanism' provided the management and organisational structure for the implementation of individual local projects. The delivery mechanism ensured the coordination of local projects and the overall supervision and direction of initiatives. Many projects under the Saltley–Small Heath Partnership were delivered collaboratively with the city housing, leisure, education and economic development departments. 'Topic group leaders' supported local projects drawn from city departments and they reported to the Area Manager and partnership board. During 1996, this interaction was an active process with the Area Manager establishing working relationships with the main city council service departments. The structure provided the partnership with a core managerial competence that had some characteristics of a centralised configuration (see chapters two and ten) to the extent that external partnerships were involved and relied upon central support. However, there was strong management board control over projects and in Saltley–Small Heath local managers operated under local authority oversight that gave an organisationally centralised quality to the board. The city council's round one bid for Saltley–Small Heath therefore reflected Birmingham City Council priorities for urban regeneration over the medium and longer terms. In the first year of the operation of the Saltley–Small Heath Partnership, City Pride provided a mechanism for the approval of local projects within the programme and it continued to provide a strategic framework for the Saltley Small Heath partnership when the prior approval requirement was later abandoned. City Pride objectives provided a coherent reference for local programmes, although in Saltley–Small Heath the distribution of responsibilities for strategy between, the city council and the GO–WM highlighted the network complexity and complicated managerial context within which the local programme operated. The local 'area' approach addressed more specific community needs and provided important policy lessons for managers working within the community that informed subsequent SRBCF programmes in Birmingham. For example, the focus upon youth in Saltley–Small Heath reflected a strong local concern about youth unemployment and the need for improved skills training. The programme played an important role in highlighting youth issues within the city through contacts with local politicians and city officials.

Risk sharing and political competition

The partnership management team in Saltley–Small Heath recognised the risks involved when carrying out a programme of this kind. The appearance of a section on risk in the partnership's documentation (Birmingham City Council, 1994) showed a move towards a risk sharing awareness in policy. Traditional concerns were commonly about the risks associated with meeting targets and programme objectives. However, the Saltley–Small Heath Partnership elaborated a wider range of risks arising from programme implementation and public agency cooperation with neighbourhood organisations and the corporate private sector. Risk analysis was 'a routine element' of urban regeneration (Birmingham City Council, 1994, p. 5) concerning the meeting of objectives in programmes that were often difficult to coordinate or where the managerial competencies of agencies varied. Risks increased where partnerships created new organisational structures and extended their activities within communities. The Saltley–Small Heath Partnership was risky because partners sought to establish an urban regeneration initiative through a process that involved 'painstaking' preparation, training and timing (Birmingham City Council, 1994, p. 5).

There were also political risks, especially for the city council, in establishing the partnership. Conflicts could hamper the efforts of the Area Board and jeopardise the meeting of policy objectives. Indeed, disagreements arose within the community about how the partnership should distribute and prioritise resources, and the Area Board had reservations about the centralisation of decision making with the city council and the GO–WM. Members of the Area Board often felt that they had little impact on policy priorities for the five year life of the partnership. One view was that central approval of the projects supported by the partnership reduced local discretion. Essentially it was the GO–WM that gave the final go ahead to local projects, and often the views of the local community were marginalised. As one community worker commented, just how far was the programme 'imposed on the Area Board?' The Area Board could approve projects but they could not simply alter projects on the priority list in the original SRBCF bid. The short life of the partnership meant that new and innovative projects could not easily be approved, so the room for flexibility was very small.

Community involvement was by way of an East Birmingham Community Forum, operating as a consultative mechanism in Saltley, and through a Small Heath Forum. These drew members mainly from local voluntary organisations, and the forums provided representatives to the Saltley–Small Heath Partnership Area Board. Early disagreements existed within the partnership about which organisations best represented local communities, although partnership managers recognised a need for more effective consultation of organisations that could influence the partnership. Demands for the representation of youth organisations reinforced the impression that 'insider' groups could influence board decisions and establish priorities that did not necessarily reflect local opinion. Public meetings enabled the Area Manager to monitor public attitudes, but many felt that the partnership should broaden its base. There had been positive community involvement in the fund bidding process, but local public dissatisfaction with past regeneration efforts in the area persisted. Poor public perceptions of central government policies and the alleged failure effectively to deliver programmes in Birmingham fostered an impression of negativity. The East Birmingham Community Forum, established by a Conservative government-created Task Force, did little to alleviate public dissatisfaction. One problem was the lack of opportunities for local groups to develop new projects, and a view shared by some local activists was that some groups were isolated despite the diversity of a community with many vocal ethnic minority groups. Some local activists detected competition between groups representing different minority interests that strongly defended their own agendas and demands.

Joined-up government?

The Birmingham research illustrates the multilevel complexity of public–private partnerships in local economic development and the tendency for urban regeneration partnerships to depend upon centralised oversight and funding. Administrative fragmentation and organisational fuzziness produced a desire among public officials for coherence and coordination. A national evaluation of early SRBCF partnerships (Brennan, Rhodes, and Tyler, 1998) for the DETR found that the SRBCF had produced better coordination and a greater awareness of strategic issues. However, the above research shows that these were often difficult to achieve across the city because programmes such as SRBCF and European Union programmes received funding at different times, operated with separate local managements and worked to different policy time scales. The city council confronted many problems in monitoring such initiatives, but it managed to provide a focus for discussion about the role of the various local projects within the city. This underscored the tendency within fragmented administrative settings for public officials to control and restrain networks by seeking centralised solutions despite the notion of 'hands-off' management (Brennan *et al.*, 1998) in SRBCF partnerships.

A third way?

There are various ways to evaluate local strategies and partnerships. These are covered in chapter ten. However, a crucial test of New Labour's approach to urban regeneration will be whether or not programmes can achieve strategic direction and also empower communities to overcome social exclusion and lack of access to institutions. Hierarchy through centralisation and the tendency of partnerships to adopt top–down solutions run counter to effective strategy and meaningful community empowerment. The Labour government's search for a third way as an alternative to both Thatcher-style neoliberalism and state socialism relies upon holistic programmes with the maximum bottom–up involvement in programmes, voluntary effort, and democratic social networks (Giddens, 1998). These may be attainable, but policy makers work in complex policy domains where networks structure inequalities at different levels and in different ways. Community involvement at one level is not necessarily matched by involvement at another level.

The evidence in this and the previous chapter suggests that community organisations in Birmingham increasingly played important roles in local partnerships and through ward level consultation, but they had limited influence over regional policies (chapter six) where they featured less in strategic statements. There was a gradual fading-out of community influence the further groups operated up organisational and spatial hierarchies in the West Midlands. Even on the Saltley–Small Heath Partnership board, where local groups had representation at the level of the local programme, they often felt that predetermined priorities provided little scope for local innovation. The GO–WM and the city council monitored the partnership in Saltley–Small Heath and had ultimate responsibility for its success. Community involvement did have influence on the partnership board's decisions about specific local interventions, but community organisations had to negotiate with the city and GO–WM through the board for any changes to objectives or fund allocations. The impact of groups thus depended on how they were incorporated in programmes and the degree to which they could influence the most important decision takers. Community organisations, to be most effective, had to work with more than one strategic apex, and in the Saltley–Small Heath example, they needed effective relationships with the GO–WM, the city council and the partnership board. With limited resources, neighbourhood organisations could expect to have little impact on more remote or inaccessible bodies, and with the Regional Development Agency assuming responsibility for the SRBCF in 1999 there were sure to be new political relationships and patterns of bargaining over resources and political access.

8
Rotterdam Scenarios

This chapter reports research carried out in Rotterdam between early 1996 and the end of 1997. It is about the development of innovative strategies and new partnerships at a time of complex contingent change influencing urban and regional policies. It deals mainly with the 'dry economy' in Rotterdam as opposed to the problems of the port and the city's 'wet economy'. The port featured in the city's strategic thinking in a big way, but the Municipal Port Management had its own departments dealing with strategic intelligence, social and economic development, regional planning, and transport and logistics. In 1998, central government took part in a review of the port management with a view to redefining its management roles and operational relationships with the city. A city council Board of Commissioners had political responsibility for the port, but there seemed to be growing support for greater management autonomy for the port and less dependence by the city on port revenues. The Main Port authority itself lobbied central government and worked in partnership with the city on planning issues. However, the following concentrates on visioning from the point of view of the city council and the attempts better to link the wet and dry economies.

While there was no simple reflection of strategy in patterns of local organisation, the city provides an illustration of how local political commitments influenced partnership building and policy implementation. Mayor Peper's involvement in the influential Eurocities network of cities provided Rotterdam with an influential voice within the European Union. Mayor Peper, who became President of Eurocities before he left to join the national government as Minister of the Interior in 1998, championed the Eurocities agenda that by 1996 focused on social inclusion and local self-governance. Rotterdam's attention to networking through Eurocities enabled the city to consolidate it's high reputation for policy innovation as well as underlining the strategic importance of the city within the wealthy heart of the European Union (Harding *et al.*, 1994).

Regionalisation

It is tempting to regard the whole of the Netherlands as a vibrant European regional economy within which Rotterdam and its World Port are vital to national competitiveness. The country lies in the prosperous economic core of the European Union, and viewed in this context the whole nation is an active economic player in the global economy. The Netherlands Bureau for Economic Policy Analysis (CPB) therefore refers to the importance of cooperation between central, provincial, and local governments to achieve national competitiveness (CPB, 1997). The Scientific Council for Government Policy, the National Spatial Planning Agency, various public agencies and government ministries all have strategies for economic development, planning and environmental policy that regard national policy initiatives as policy priorities.

Toonen (1998) argues that in this context 'regionalisation' in the Netherlands has never been clearly articulated as 'regionalism'. Regionalism implies strong regional public institutions and policy direction, whereas the debate in the Netherlands has been more about partial administrative reforms and functional reorganisation. For Toonen, the 1990s ushered a period of profound change in Dutch politics as the traditional foundations of the politics of accommodation changed to produce a reordering of regional politics. However, pillarisation (chapter one) had produced socially and politically integrative institutions that were functional in supporting the effective political management of regional policies. Regional disparities were not important politically even though the pillarised parties and social groups related to different geographical areas in the Netherlands (Toonen, 1998). Pillarisation served to overcome the political conflicts and tensions between regions through a political system that integrated regional subcultures and minority interests into the wider national community thereby 'nationalizing' regional interests. Toonen (1998) argues that this explains why constitutional regional government was absent in the Netherlands. In the modern world of policy networking, the 'region' was not fixed, but was an all-embracing concept used by different 'branches' of Dutch public administration each with their own regional spatial definitions. According to Toonen, the individual Dutch provinces came closest to what a foreign observer might label a 'region', but Dutch administrators regarded provinces as the least important part of the administrative system. Provincial boundaries did not conform to actual patterns of economic activity and there was administrative ambiguity in regional–provincial level institutions. Toonen refers to the 'gewesten' as an administrative level between the provinces and municipalities consisting of intermunicipal co-operative districts, and central government 'field' organisations. The gewesten, and the comparatively small spatial coverage of the provinces, increased the difficulty of defining a coherent regionalism in the Netherlands (Toonen, 1998, p. 137). In the 1990s therefore, complex networks of organisations linked spatial levels so that local governments cooperated in service delivery, economic development, and social policy in a 'multiple meso' (Toonen, 1998, p. 142).

The Randstad

The Randstad (Rim City) illustrates the complexity of Dutch regional governance. The Randstad includes the provinces of North Holland, South Holland, Utrecht, and Flevoland (see Introduction). According to the Randstad Cooperation–Economic Affairs (ROEZ) organisation, the Randstad covers an area of 6000 square kilometers and forms an urban ring in the heart of the Netherlands with a 1994 population of 7.1 million (ROEZ, 1996). The ROEZ views the Randstad as a powerhouse of economic activity and a concentration of cultural and social diversity. Dileleman and Musterd (1992) trace the growing economic influence of the Randstad after the Second World War when politicians and planners pursued urban development around the 'Green Heart' of the country. The ring concept appealed to urban planners because of its conceptual neatness and because it represented the dynamic centre of the expanding Dutch economy. The Randstad cities faced common problems of economic and social development, so viewing their destinies as a whole was appropriate. Dileleman and Musterd refer to the need for social housing, land for industry and the desire to control urban expansion which led planners to view economic development in regional terms. This enabled central government to establish national priorities for urban and regional economic development with a variety of agencies implementing central government programmes. However, by the 1990s Dileleman and Musterd (1992, p.14) could comment that 'the municipalities no longer coincide with the functional entities in which urban life takes place. Neither the housing market nor the labour markets operate locally, but rather at a regional scale.' By the 1990s, the lack of coherence between administrative entities and real economic activity forced planners to adopt multiple policy foci that accounted for the clustering of economic activities around two especially active economic areas within the Randstad. The north wing of the Randstad included Haarlem, Amsterdam and Utrecht, and the south wing extended from Leiden down to the two central cities of The Hague and Rotterdam. The cities within the south wing influenced each other economically (City of Rotterdam, 1992), and urban development between them created new corridors of development in a regional concentration of 2.5 million people. Planners thus regarded Rotterdam at the hub of two economic zones; one linking northwards to the north wing of the Randstad, and the other linking the city to the south along Highway A15 to West Brabant and Antwerp (City of Rotterdam, 1992). The more narrowly defined Rotterdam urban region, including the city and its immediately surrounding municipalities, had a population of over one million and strong economic competencies in port activities, business, commerce and services.

Van der Wusten and Faludi (1992) discuss the administrative implications of the Randstad. They refer to a Dutch planning system complicated by many agencies, territorial boundaries and responsibilities affecting public and other organisations. They argue that administrative complexities strongly influenced physical planning after the Second World War and that it was necessary for agencies to work together to assess national policy priorities in planning and economic development. Cooperation between municipalities reflected the need to work across jurisdictional boundaries, although municipalities were sometimes slow to identify common interests. In the 1990s, rivalries were evident as cities each tried to enhance their status and looked beyond 'functional urban regions' based on commuting and service activities.

Interagency cooperation and interprovincial cooperation in the Randstad led to new public policy strategies. Kreukels (1992) refers to the development of effective administrative arrangements in the Randstad, arguing that despite the strong influence of central government in the control of planning it was important to preserve decentralised regulation of the physical planning regime. This placed responsibilities on authorities at the regional and local levels for citizen involvement, strategy and the control of urbanisation. For Kreukels (1992, p. 242), the Dutch system's unique characteristics influenced the physical planning process producing 'social involvement, professionalisation, government interference, and a strong emphasis on the national and provincial policy efforts in the post Second World War period. All of these factors appear to distinguish the Netherlands from other countries when it comes to physical planning.' The 1970s was the 'golden age of physical planning' (Kreukels, 1992, p. 243) after which policies focused on different priorities. Private sector involvement, the market, and decentralised solutions became more important. By the late 1980s and early 1990s, the effort to 'counter the highly bureaucratic and centralistic government interference' (Kreukels, 1992, p. 243) gathered momentum as market policies spread to local government and as local interest groups demanded a greater say in policy. For Kreukels, the policies that combined citizen involvement and market objectives inevitably challenged the bureaucratic and government 'establishment' (Kreukels, 1992, p. 243). However, Kreukels argues that national planners failed to recognise the tensions between conflicting interests thus leaving the fundamental issue unresolved concerning how the market could effectively be accommodated and developed within the changing Dutch political system. This helps to explain the tensions that remain in Dutch economic development policies between the state and the market, between city governments and the private sector, and between bureaucrats and citizens. Therefore, policy still embodies elements of the old hierarchical planning mind frame as well as elements that encompass the market.

Regional political competition

The persistence of old cultural assumptions might in part explain why Dutch local government reform has not kept pace with the dynamic economic change going on in the economy. For example, Andeweg and Irwin (1993) see little scope for an enhanced role for provincial governments in strategic planing in the Netherlands because the provinces have not satisfactorily adapted to changing conditions. Provincial governments are directly elected, but they are regarded by many public officials and politicians as ineffectual. As Andeweg and Irwin (1993, p.159) argue, 'the big cities of Amsterdam, Rotterdam,

The Hague, and Utrecht often view the provinces as unwelcome representatives of the small municipalities that surround them, and hinder their expansion'. However, adjacent to Rotterdam in the economically active province of North Brabant an active development agency funded by central government and the province provides investment support to companies and collaborates with innovative companies to strengthen competitiveness and market share.

Such involvements mark a recognition at provincial level that there are pressing economic issues. Nevertheless, a study for Amsterdam City Council (1994) cites the concern of many public officials with the slowness of administrative reform and the underlying weaknesses of provincial administration in economic development. But sluggish administrative reform also affected the big cities. In Rotterdam, where despite intermunicipal cooperation through the Rijnmond Intermunicipal Agency (1994), a proposal to create a metropolitan 'city province' was obstructed by political controversy and inertia (Hendriks and Toonen, 1995). Moreover, the Randstad as a whole failed to provide the conditions for new regional governance in the 1990s. This is partly because it is not socially or economically cohesive in spite of the apparent neatness of its geography. Indeed, Randstad cities differ widely in character and they defend their own interests and traditions. Therefore, regional politics are played out in a variety of networks. The Randstad Joint Authority Arrangement is a collaborative network that links the provinces in the region, and the four big cities of Amsterdam, Rotterdam, The Hague and Utrecht jointly constitute an influential lobby on issues such as housing, economic development, and transport. A municipal network (VNG) relies on diverse contacts between the big Dutch cities and central government, and is generally regarded as more important than the joint authority arrangement. The big cities lobby government through the office of the Prime Minister and consultation takes many forms depending upon the policy issues involved. Other lobby groups include chambers of commerce, agricultural interests and the provinces. Rotterdam City Council is active in many networks and has demonstrated a concern over a range of regional issues including the location of businesses away from the city on 'green-field' sites, the formulation of a Netherlands 2030 policy for planning with central government and related spatial planning issues.

The tendency to stress local interests is reflected in the perception in Rotterdam that there is a north–south issue that influences development in the Randstad. According to this view, in the north Amsterdam leads development and is a major international point of attraction for transnational corporations and inward investment. The area around Schiphol international airport is an economic growth pole that is extending to the southern part of Amsterdam and creating an important European cluster of high technology, finance and services. Rotterdam takes an interest in this economic development, but wants a share of the action further south. So, strategy in Rotterdam has to take account of the competitive environment, the future of the port, and the role that the city will play in the economic development of the Randstad. These issues underline the competition between the cities in the Randstad as well as their collective cooperative interests. Regional political competition forces cities to strengthen their individual strategic policies and to focus upon the competencies of their respective metropolitan regions. This adds to the difficulties for Rotterdam such as the political problems of creating a metropolitan authority, and Toonen (1998) asserts that the lack of agreement in the Rotterdam region reflects a deep crisis of administration. This was at just the time that the city needed to compete with other Dutch and European cities and modernise its government system in line with the demands of the time. Many policy makers therefore regarded the situation as unfortunate although the Dutch government in 1998 agreed to extend the existing regional cooperation agreement between municipalities with an evaluation of its impact in 2002. This evaluation would form the basis for further discussion about regional administrative reform.

Reflecting its self-interest, Rotterdam trades the image of the modern thrusting port city. It contrasts with other Dutch cities with regard to architectural style and the general pace of city life. Rotterdam has a different feel to it than either Amsterdam or The Hague. However, the reform of local government has been tempered in response to local traditions and cultural expectations. Map 9.1 shows greater Rotterdam where the 'city province' proposal was for a new Dutch metropolitan authority. However, when planners promoted the city province, they were stalled by local interests that wished to keep local government close to the people. Figure 8.1 shows the populations of the small municipalities around Rotterdam. Although localist sentiments persisted there, it was the desire of Rotterdam city voters to retain the city's identity that aborted the proposed reform in a 1995 referendum.

Toonen refers to the strength of local political interests and the

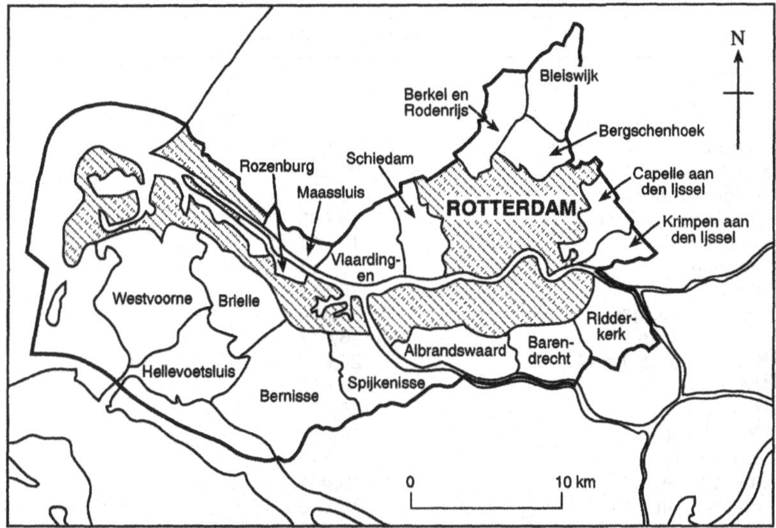

Map 8.1 Metropolitan Rotterdam

protracted debate about Dutch regional government in the 1990s. The 'Rotterdam model' led central government to pursue its city province policy nationally (Toonen, 1998, p. 132) and it envisaged the formation of a metropolitan authority for the Rijnmond, or 'greater Rotterdam'. The history of local government reform in Rotterdam highlights the Dutch capacity for extensive consultation, debate and political bargaining. While positive administrative changes have been implemented, the record has also been one of failed experiments, frustrated ambitions and bureaucratic wrangles. The defeat of the reform of regional government in Rotterdam thus accompanied the development of informal networking as the city council expanded its contacts with surrounding jurisdictions, corporations and nonprofit organisations as the city collaborated regionally with provincial authorities, the national government and a wide range of corporate investors eager to invest in the city. Local administrative reform in the 1990s encouraged new forms of public service provision and delivery, and management processes in city and provincial governments were improved as economies and efficiencies were achieved. However, in 1995 when the referenda in both Amsterdam and Rotterdam went against city province proposals, central government had to rethink its plans by initiating further

Figure 8.1 Populations of Rotterdam and Surrounding Municipalities, 1994

Area	Population (Estimated)
Rotterdam	598,521
Municipalities:	
Albrandswaard	14,920
Barendrecht	21,317
Bergschenhoek	8,058
Berkel en Rodenrijs	15,663
Bernisse	12,267
Bleiswijk	9,617
Brielle	15,514
Capelle a.d. Ijssel	59,364
Hellevoetsluis	36,617
Krimpen a. d. Ijssel	27,627
Maashuis	33,055
Ridderkerk	46,425
Rozenburg	14,204
Schiedam	72,515
Spijkenisse	70,464
Vlaardingen	73,820
Westvoorne	13,780
Total Metro Area	1,143,648
Ethnic Minorities in Rotterdam City (excluding surrounding municipalities)	
Surinamese	45,724
Turkish	34,916
Moroccan	25,433
Southern European	15,632
Cape Verdean	13,073
Antillian	11,398

Sources: Rotterdam City Council and CBS (from different data sets).

consultations with local governments and political parties. Central government pondered the future of the larger city regions, but even by early 1997 little progress had been made in reconciling conflicting regional and local interests. Despite attempts to arrive at an all party consensus, city provinces in Rotterdam and six other metro regions could not become reality until there was agreement on the shape of metropolitan government and the relationship of the cities to surrounding authorities. But, by 1998 the city province project had collapsed nationally.

Strategy and partnership

All this increased the uncertainties of local government officials and politicians about the future direction of economic and social development. At the time of the research conducted between 1996 and early 1997, Rotterdam was governed by Mayor Bram Peper and a multiparty group of Aldermen representing the Labour party, Greens, Liberals (VVD), Democrats (D66) and Christian Democrats (CDA). The broad recognition by the Mayor and Aldermen that Rotterdam and its port were facing intense competition in world markets reflected a consensual approach to policy. Well before the 1995 referendum defeat of the city province proposal, the city had developed a range of policy initiatives that involved neighbourhood and private sector groups in urban regeneration, and in the mid 1990s, this approach was redefined and focused through extensive multiparty consultation and consent. The city council adopted the 'coproduction' of policy (Anderiesse, Bol, Oudijk, and Bons, 1997, p. 20) that involved the council in extensive consultation with community and business groups.

Central to this approach was the Rotterdam City Development Corporation, or the Ontwikkelings Bedrijf Rotterdam (OBR). The OBR, created by the city council, had an interest in the city's 'dry economy', as opposed to the 'wet economy'. The OBR facilitated sites for economic development and housing in the city, and was involved in several regional development projects. The OBR assisted housing projects in the municipalities surrounding Rotterdam, and it had expertise and financial resources that enabled it to lend professional assistance to other organisations. As a risk-taking organisation, OBR invested money in land and infrastructure, commissioned commercial construction projects and developed a strong project management capability. It became what one official described as 'a strategic organization' committed to policy innovation and diversification in the services that it provided. The OBR therefore pursued a proactive strategy in a changing external environment where public organisations had to be as flexible as the best competitive companies. However, OBR officials recognised that innovative policies did not automatically produce organisational flexibility, and in many ways the OBR lagged behind its own ambitious policy aspirations. Traditional institutional constraints sometimes impeded change which produced a tension between policy innovation and the public service ethos.

Political restraint and business

Consensual politics consolidated political restraint through OBR's extensive intergroup collaboration and public consultation. The OBR supported partnerships linking communities with the private sector through the coproduction policy, although business involvement in neighbourhoods was limited compared to the USA (Anderiesse *et al.* 1997). However, the Rotterdam Chamber of Commerce articulated business interests within the city by representing members and by adopting a strong regional orientation. The chamber worked closely with the city council on regional policies (Rotterdam Department of Urban Planning and Housing, 1992) with strategic initiatives for infrastructure and housing, and a multiagency initiative for regional development. The chamber maintained contacts with a Regional Economic Organisation, for which the chamber acted as the secretariat, and chamber members informally networked with the Mayor and Aldermen and influential regional policy makers. Chamber representatives interviewed for this research reported 'excellent' working relationships at all levels with city officials and the port authority where there were close corporate and public sector relations in economic affairs.

The city council and the chamber did not always share the same goals, and chamber officials referred to both agreements and disagreements between private and public sector objectives. However, while conflicts with the city arose over economic policies, chamber officials maintained that the relationship with the council was one that produced mutual benefits and partnership. The chamber's affiliated companies lobbied to promote their interests concerning commercial incentives, spatial policy and the improvement of Rotterdam's infrastructure. This produced a strong incentive for the public and private sectors to present a common view when dealing with central government departments, and it provided chamber members with the chance to approach city officials when major developments were proposed. However, the chamber's role was legally constrained to the extent that it was not permitted to make direct speculative investments in projects. The chamber thus acted as a facilitator and policy networker for companies so that they could benefit from commercial opportunities.

The Rotterdam Port Promotion Council was another forum for business, promoting the port through a public–private partnership between Municipal Port Management and companies in the greater Rotterdam region. With three hundred member companies, the council held public relations and marketing events and supported business networking with a membership including P & O Nedloyd, Rotterdam Airport BV, Coopers and Lybrand, Shell, ING Bank, KPMG,

ABN AMRO Bank and Deloitte & Touche. These companies especially retained an active interest in the performance of the World Port as an important aspect of their own commitment to the city.

Policy innovation: dealing with complex contingencies

Informal corporate networks, and the relationship of large corporations to the Port of Rotterdam Authority, received little public attention despite the policy of coproduction. Given OBR's strategy, Rotterdam needed to assess its role globally, regionally, and nationally and a positive attitude toward the private sector was crucially important. However, the OBR and city council also needed to develop an effective relationship with local people who would participate in the future development of the city. Social and economic development were therefore important priorities in linking business with neighbourhoods through coproduction. Social and economic policies were also set within the framework of national policies, although the city had significant discretion over the formulation of local and regional programme objectives. Officials at OBR described the relationship between the city council and central government as a close one where the priorities of the city were taken into account despite occasional differences. This cooperative approach was essential in economic development where, for example, policies involved the city with the Ministry of Housing, Physical Planning, and the Environment, and the Ministry of Economic Affairs.

Scenarios

An important part of the city's strategic planning effort in the mid-1990s was the development of forward planning to account for the complexity of the management and economic development tasks facing the city. Through its Innovation Unit, the OBR stimulated new thinking on the possible future courses for the city and the major economic growth corridors around Rotterdam and Antwerp. Officials at OBR, as part of this exercise, wanted to examine the future relationship between Antwerp and Rotterdam and the competition with other cities as well. This was important given the OBR's perspective of Rotterdam as part of a worldwide economic system where corporate investments would divide between competitive cities. International companies did not only invest in manufacturing and distribution facilities, they also shifted custom for services between cities and moved goods between different ports.

Competitive cities vied for international trade and inward investment, and OBR officials recognised that the jurisdictions of Dutch local, provincial and national public organisations did not match new patterns of economic activity. The city thus had to broaden its policy perspective taking account of dynamic economic trends within the global economy and expanding regional growth nodes.

To help with the process of discussing the possible paths for regional economic and community development, the city council appointed the Global Business Network (GBN) as consultants. The GBN drew on management techniques developed in the 1960s by Pierre Wack at Shell, and advocated developing strategic scenarios that could be used in public organisations and local governments. Van der Heijden (1996) describes the GBN method by distinguishing scenario planning from Mintzberg's (1994) intuitive strategy and Porter's (1985) optimal strategy approaches. For van der Heijden, scenarios enable planners to enter strategic 'conversations' that are not possible in the Mintzberg and Porter models. According to van der Heijden, Mintzberg disempowers managers because the strategy process is portrayed as ambiguous and unpredictable. Porter assumes that managers are rational actors, but for van der Heijden rationality does not always work. What may appear to be perfectly rational and acceptable one day, may not be appropriate in the future, so managers should engage in conversations about a range of possible future outcomes. While the conversations require formal management involvement, they also involve informal discussions about possible outcomes. Managers have to be aware of the vocabulary of the conversations so that they can communicate ideas effectively and question long-held policy commitments.

Van der Heijden describes scenarios relying upon a synthesis of different organisational strategy perspectives. An important aspect of the scenario approach is that it encourages organisational learning that integrates 'action and experience in the strategy development activity' (van der Heijden, 1996, p. 55). Different organisations learn in different ways and they view the external environment through filters reflecting their internal commitments and values. Organisations must break down preconceptions and overcome biases that hinder a proper perception of reality. Normally organisations ignore important changes because policy makers are preoccupied with 'events that catch the attention' (van der Heijden, 1996, p. 55) while less dramatic, but more fundamental, changes escape immediate view. Policy makers tend to deal with problems that most affect their own interests which, according to van der Heijden (1996, p. 56), can lead to an excessive preoccupation with 'a struggle for survival' at the expense of a longer view of an organisation's development. Instead of short-termism, policy makers need a 'business idea' to drive organisational thinking and strategy. The scenarios developed in this process of organisational thinking build the business idea through an assessment of the possible futures facing policy makers. External environmental variables can be responded to in different ways, so the scenarios provide what van der Heijden (1996, p. 57) calls a 'strategy test bed'. Sometimes a scenario may produce a final decision, 'but normally the manager uses the scenarios to test strategy proposals in order to find ways to improve them, i.e., make them more appropriate and robust against the futures that might arise' (van der Heijden, 1996, p. 57).

In Rotterdam, scenarios (OBR, 1996) enabled policy makers to think about the future in new ways. The 1995 exercise involved council members, officials, education organisations, representatives of government ministries, companies, and community groups organised into workshops to discuss several scenarios and their policy implications for the year 2015. Some of the politicians were skeptical about the process and felt that the exercise was divorced from reality because the scenarios were presented abstractly. However, gradually these reservations gave way to support for the process and OBR and senior public officials used the scenarios as points of reference in the development of the city's economic and social development strategies (OBR, 1996). Policy initiatives, such as Agenda 2000, thus addressed the lessons learned from the scenario process.

The following account was informed by Kees Machielse of the OBR who was a prominent member of the Scenario Project Team, and by Japp Leemhuis of the Global Business Network who advised on the organisational aspects of the Rotterdam exercise. The four scenarios outlined below depict Rotterdam 'Worldwide,' Rotterdam 'shackled', Rotterdam 'uncoupled', and Rotterdam 'talented'. Figures 8.2–8.4 show summaries of the scenarios based on translations from the Dutch, but adapted and compressed to aid exposition. However, the figures contain the important assumptions in the original scenarios that showed trends in social and economic conditions that could combine to produce policy problems for the city. It was likely that in reality some features found in each of the individual scenarios would be found together, and in this way the approach had similar implications to those in cultural theory (chapter one)

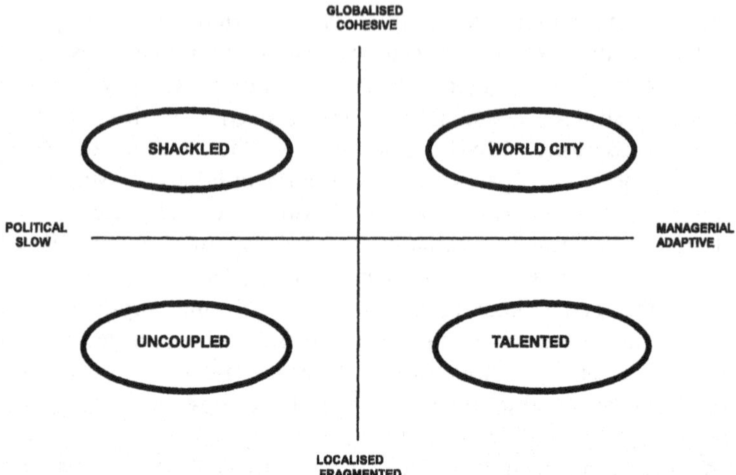

Figure 8.2 Rotterdam: The Scenario Diagram (with permission from OBR)

where subtle interactions and cultural outcomes reflecting multiple influences were likely.

Figure 8.2 shows the basic features of four scenarios derived by measuring degrees of government fragmentation along the vertical axis, and government capacity to respond to events on the horizontal axis. The policy responses of the city council are likely either to be effective, or very slow and bureaucratically constrained. The plotting of Rotterdam's characteristics on points along the axes produced combinations of possible circumstances that defined particular scenarios. So, in the Rotterdam shackled scenario, the city is trapped in a situation where policy responses to the external environment are conditioned by managerial slowness and bureaucratic and hierarchical incapacity. Figure 8.3 provides an overview of the scenarios covering aspects of the city's social and economic performance. The port–city relationship is clearly problematic in the scenarios, but perhaps the most revealing aspect of Figure 8.3 is the item dealing with the typical image of the city. The prospects include that of a logistic city like Singapore or, at worst that of Moscow. Figure 8.4 summarises the international orientations of the city under each of the scenarios.

Rotterdam Worldwide envisages the development of Rotterdam as a World City. The 'World City' is a vibrant and prosperous city where citizens enjoy all the benefits of the economic, social and cultural

Figure 8.3 Rotterdam Scenarios: Overview

	Rotterdam Worldwide	Rotterdam Shackled	Rotterdam Uncoupled	Rotterdam Talented
Ability and capacity	Renewal & innovation	Low dynamic	Self-regulating dynamic	Natural dynamic
Value-creation	Large-scale alliances	Poor back-up for alliances	Local economy is driving force	Small-scale attitude & specialisation
Type of economy	Part of world economy	Sector maintenance and emergence of small informal economy	Informal street economy-barter as well as formal economy	Directed to services and small-scale production
Unemployment	Low unemployment with social safety net	Once high, but many alternative jobs	Low, because informal economy employs	Low unemployment with varied labour supply
Harbour-city relationship	City and harbour well integrated	Harbour not well integrated	Smaller harbour with local and regional focus	Smaller harbour, with dominant city
Education & knowledge	Knowledge is essential. Permanent schooling for people to keep up-to-date	Emphasis on allocating educational resources	Different groups have different degrees of access	Crucial for everybody
Environment	Environment sector is a new growth market with technological solutions	Major environmental issues neglected	Narrow perceptions of environmental issues	Policies aimed at durability
Typical image Example	Logistic city Singapore	City is 'stuck' Liverpool	'Third-world' city Moscow	Well-balanced city San Francisco

Source: Adapted with permission from OBR, 1996.

Figure 8.4 Rotterdam Scenarios: International Orientation

Rotterdam Worldwide	Rotterdam Shackled	Rotterdam Uncoupled	Rotterdam Talented
Policy concerns in Rotterdam begin to focus on the international level. Under the umbrella of large corporate conglomerates, smaller firms in the city operate internationally. This is especially evident in business management services and the construction services industry. However, more foreign companies become active in the domestic market and foreign products and services appear in the market in Rotterdam. This applies to industrial products and transport.	Rotterdam has space problems. Expanding firms that are not bound to the region are forced to move to locations outside the Rotterdam region. Activities presently based in the region are agriculture, foodstuffs, and building. Transport and wholesaling are also important. Services relate to these activities, but they too will be mobile. Consequently, the departure of major internationally operating concerns would harm the regional economy and reduce the income of the population.	Companies that operate internationally face competition and sometimes go out of business. In contrast to the 'talented' scenario, no alternative develops. The local economy does not generate adequate alternatives, although there could be growth in the informal economy.	It is almost certain in this scenario that in the early stages business will operate internationally. However, there will be intense international competition. There is a high dependency in the region on certain sectors such as the chemical industry, transportation, financial services, and agriculture; all of which come under pressure. Some companies will go bankrupt; others will contract and re-orientate to the domestic market. This will take time to happen. However, firms do not relocate because conditions are poor elsewhere.

Source: Adapted with permission from OBR, 1996.

dynamism associated with growth and prosperity. Despite the problems of poor communities, social tensions, racism, and social inequality, the largest companies invest in the World City, the best orchestras play there and the most exciting buildings are built there. The World City evokes multiple images of the social life of citizens who enjoy access to international culture, cuisine and social intercourse. The World City is 'the place to be', and it is unpredictable and exciting. World City status is not something that politicians and planners can necessarily achieve by intent. The great world cities such as London, Tokyo and Paris were not created as international cities by setting policy aims and objectives. They developed as the global market changed, as their industries and services strategically positioned internationally, and as their cultural attractions appealed beyond national borders. The world city develops because there is a market for its goods and services and because the conditions are right for there to be a synergy between its different activities.

The scenario presents Rotterdam developing along its own unique path, but policy makers and business leaders can influence the ways in which the city grows even though they cannot be sure that it will ascend the global urban hierarchy. The scenario envisages the city and its region as playing a bigger part in the world economy with Rotterdam based companies entering global alliances in industry, commerce and finance. The port plays an important part in the Rotterdam region's success with its role as World Port continuing to make the city a hub for trade and exchange. According to the scenario, the harbour and the city develop a close relationship where policy makers ensure the integrated economic development of the two, and the city benefits from the port as the port grows from the region's strength. The focus of public policy is innovation and the maintenance of a strong market economy as high technology companies benefit form an education system that concentrates on new applications within information based industries. Unemployment declines because the city develops an enterprise economy geared to growth and the full development of its human resources.

Rotterdam shackled presents a different picture that some politicians and officials feared might best describe the actual trajectory of the city. In this scenario, civic leaders become complacent because they rely on past success formulas. New challenges require new strategies and policies as the market changes and competition intensifies. Resting content with existing policies may be sustainable for a while, but as conditions change competitors gain the advantage.

Rotterdam's leading position as World Port is not guaranteed, and the city is 'shackled' through satisfaction with things as they are, or as they have been. The city is 'trapped' because policy makers and business leaders fail to set an alternative course to remain competitive. Gradually, global competition undercuts the city, the port, and the region, and a fatalist culture emerges (see chapter one) as international companies look elsewhere for an entrepreneurial and innovative environment within which to work. According to OBR, at worst this pictures Rotterdam as a potential Liverpool of the Netherlands; a city 'stuck' with a developing informal economy and cuts in local services and educational provision.

Rotterdam uncoupled presents the city failing to sustain its economic growth and losing market share globally; hence there is a danger that the city will become 'uncoupled'. This is a protracted urban crisis in which social and economic conditions produce policy problems that lead politicians and public officials to fight on various fronts. They feel pressured as the city experiences acute social tensions and declining morale in the business community. There is the prospect of conflicts between different social groups, racism and social unrest.

Rotterdam talented presents quite a pleasant prospect by contrast. According to this scenario, the city may not achieve the status of a World City, but its inhabitants gain from the flourishing of the talents and inventiveness of local people. In the era of information technology, Rotterdam develops new competencies by using new technologies to expand the opportunities of its citizens. The 'digital city' can empower citizens by bringing them in contact with each other and by producing new social networks and commercial opportunities. Rotterdam talented is a civilised city in which people matter and where there emerges a confident and possibly more egalitarian culture. The scenario likens the future of Rotterdam to San Francisco with its multicultural population, cultural vitality, and active and prosperous local communities.

Policy and organisation

Translating the lessons of the scenarios into practical public policy and positive cultural change was difficult especially given the interplay between different community interests and the relations between central and city government. Public officials in Rotterdam often reviewed social, economic and administrative policies and stressed the importance of public–private partnerships in achieving global competitiveness. Economic development was a central theme in the New Rotterdam initiative which, by 1988, included flagship projects for social and administrative renewal. Projects featured local neighbourhoods as partners in community development thereby setting a precedent for the 1990s with mechanisms to improve the delivery of local services. New Rotterdam enabled the city council to consolidate areas of good practice, but by the mid-1990s it was necessary to reassess the initiative and rethink policies with the aid of the scenarios and lessons from past good practice.

The original New Rotterdam policy reflected many of the traditions of the city's former top–down planning regime with strong policy direction in social and economic renewal. However, by the mid 1990s the emphasis on private sector and nonprofit involvement required greater emphasis on multiorganisational partnerships and administrative structures to extend Rotterdam's regional interests. Social, economic and administrative renewal were therefore related, as in the early New Rotterdam policy, but in the 1990s relationships between policy areas had to be redefined holistically. According to one council official, the revised policy therefore 'unwound the social, economic, and administrative areas of policy in order to reintegrate them on a new basis within a new policy perspective.'

Agenda 2000, published in October 1995, defined the integrative approach. Agenda 2000 resulted from discussions between politicians who spoke of the need to integrate social, economic and administrative renewal policies within the new framework. The policy embodied the good practice of the original New Rotterdam, but better accounted for the competitive market where public–private partnerships needed structures suited to rapidly changing conditions. Integrating mechanisms linking social, economic and administrative processes figured prominently because underlying structural economic changes were transforming the relationships between business, local government and local community groups. By 1996, Agenda 2000 drew on the scenarios and real policy experiences and lessons to foster new organisational flexibilities and to speed innovation. Officials saw this in broad strategic terms taking account of the new roles of the public sector working in conjunction with the private sector and community organisations. Agenda 2000 thus signified an important move in the direction of opening up the policy process to private sector and nonprofit interests and away from top–down planning. Public consultation in Autumn 1996 led to the refinement of policy in the *New Rotterdam, Course 2005* that combined the new economic and social renewal approaches by linking them to the central government funded Major Cities Policy (chapter nine).

Fresh thinking

The scenarios informed policy, but it is difficult to quantify their impact. What they did was to enable public officials and politicians to examine the problems of the city in a unique and stimulating way. The scenario exercise generated ideas about the possible futures that citizens could face and about external trends and threats. Of course, the scenarios could become self-fulfilling prophecies, blinding policy makers to certain policies and strategies. As Mintzberg (1994) suggests, scenarios often serve to reinforce the prejudices of policy planners by structuring futures to fit preconceived strategies. Policy makers can be preoccupied with scenarios that undermine innovation by leading to over optimistic expectations about the future. However, the four Rotterdam scenarios did not provide fully formed visions of the future or justifications for existing policies. Public officials did not regard them as rigidly defined alternatives for the city. Instead, the scenarios increased awareness of issues that might otherwise have been neglected and they provided a fresh understanding of the external environment as complex and ever changing. New policy priorities were also set within the context of the reduction in Rotterdam's public servant workforce that in 1996 stood at 18,000 in contrast to 30,000 ten years before. The number was set to decline with extended reorganisation in local services so it was especially important to develop best administrative and service delivery practices in economic development and urban regeneration. These had to allow for the intricate networking and partnership between a managerially competent city council, private service providers, and community professionals (Anderiesse *et al.*, 1997). Given such conditions, and the complexities of policy in general, the scenarios at least provoked thinking about desired outcomes. They raised the need for clear objectives to reduce risks in particular policy areas such as competitiveness, social exclusion and communal relations.

9
Rotterdam: Partnership and Community

This chapter shows how the policies for social renewal and local economic development in Rotterdam support restrained political competition through public–private partnerships organising local interests to support growth-oriented policies under changing market conditions that require the city to attract corporate investment. The chapter, based on research carried out between early 1996 and the end of 1997, views the Kop van Zuid Project and the Major Cities Policy (GSB). The Kop van Zuid Project aims to boost Rotterdam's international standing by attracting major corporations to the city while the Major Cities Policy targets run-down neighbourhoods and ethnic minorities. The GSB policy concentrates on community safety, quality of life, and education and employment. The city council has acknowledged serious urban problems concentrated in the older city districts along the river Maas where high unemployment persists despite the impressive growth of the city's central business district. The danger of what the city development corporation described as 'a divided city' (OBR, 1994, p. 5) will remain if social conditions in deprived communities are allowed to deteriorate. 'Social renewal' thus compliments economic development, forming a broad socioeconomic intervention at the local level integrating community groups into the development process.

A risky environment

National policies identify the largest Dutch cities as part of wider regional and urban structures, and major infrastructure and development projects are set in this framework. The concept of the 'Major City' locates cities within the 'functional urban region' that consists of a main central city and its surrounding area (van den Berg, Braun, and van der Meer, 1997, p. 2). These regions contribute economically to the national and international economies. The 'major four' of The Hague, Amsterdam, Rotterdam and Utrecht (known as G–4) provided the initial central focus in the government's Major Cities Policy, and they were later joined in the initiative by other 'urban nodes' and towns functional in the wider regional context. Van den Berg *et al.* (1997) show that the administration of urban policy in the Netherlands divides between central government ministries. These include the Ministries of the Interior; Housing, Physical Planning and the Environment; Economic Affairs; Social Affairs and Employment; and Transport, Public Works and Water Management. The Ministry of the Interior has responsibility for the centrally funded Major Cities Policy, but the other ministries have interests in the programme and contribute to other initiatives in economic development and public safety. Van den Berg *et al.* (1997) show central ministries contributing to Dutch urban policy by promoting a blend of fairness and consensus. With the decentralisation of policy (chapter one), there had been a shift in emphasis to allow municipalities to develop their own initiatives and involve local communities more in urban policy. Rotterdam policy makers wanted stable political and market conditions to attract overseas investors that wanted assurances that the labour force is loyal and social conditions favourable. The scenarios in the previous chapter highlighted the concern in Rotterdam with the prospect of social tensions in communities worsening either as the result of slow growth or too rapid growth. Conflicts could sharpen social divisions, create instability and increase political risks. Community organisations in Rotterdam enjoyed a favourable relationship with the city council, but the activities of far right extremist political organisations created uncertainty in communities with ethnic minority populations. Since the traditional approach in the Netherlands encouraged compromise and consensus in local politics, local governments provided extensive opportunities for involvement in policies. Rotterdam city council thus valued the interplay of opinions and the involvement of a wide a range of organisations in the political process which politically incorporated groups. Many groups attempted to influence the city council and sometimes they gained representation on city and district councils. Yet, as Bons (1994) shows, ethnic minorities were poorly represented in neighbourhood programmes and they suffered from a low rate of participation in community initiatives. This was despite the evident urgency of the problems faced by groups such as the Surinamers, Turks and Moroccans who had settled in Rotterdam. Bons studied the districts of Feijenoord, Nieuwe Westen and Hillegersberg-Schiebroek where he found that ethnic minorities had a low motivation that restricted their involvement in neighbourhood initiatives. This was in spite of a clearly expressed willingness by the groups to assist public officials if they could. A Participation of Ethnic Minorities Project was established to overcome the problem by organising groups to get involved in local affairs and increase their capacities for self improvement, but often progress was slow.

The Kop van Zuid: political restraint and partnership

Even in major infrastructure and development initiatives, policy makers have been aware of the need for projects that provide for business interests and a concern for local communities. However, reconciling community and business interests has often been difficult especially when so much of the policy discourse has been about civic boosterism and prestige developments. The Kop van Zuid (South End) project illustrates how the city used public–private partnerships and social renewal to foster economic development and growth. The project has been important in economic development projecting the city as a leading hub for international inward investment and it was significant, not only because of its sheer size, but also because of its potential economic contribution to the future development of the regional economy and its thematic policy emphasis on the market and business opportunity. The Ministry of Housing, Physical Planning, and the Environment forecast that the Kop van Zuid would attract international business to a new Rotterdam business district and cater for the needs of international business and finance, and it would overcome the shortage of land for business development. The project provided an important element in a vision that projected a World City type of image for the south side. Following the national Fourth Report on Physical Planning in 1988, the government designated five projects, including the Kop van Zuid, as models for the implementation of the public–private partnership approach with collaboration between the corporate private sector, the city council and community organisations. The strategy placed heavy emphasis upon the involvement of international property companies, corporate investors, and world renowned architects in the development of substantial projects.

The 10,540 hectares Kop van Zuid project area, located south of the River Maas, overlooked Rotterdam's existing central business district on the north side. The south side project area comprised of former dock facilities and land around the Binnenhaven, Spoorweghaven, Entrepothaven, Rijnhaven and Wilhelmina harbours. The area had prospered with the southern expansion of the port during years of rapid economic expansion in the eighteenth and nineteenth centuries and major harbour improvements brought a rail connection over the river completed in 1877. As in other major port cities, the demands of twentieth century technological change such as containerisation and the increased size of ships shifted economic development away from traditional port locations. Dock facilities became obsolete as port operators introduced new methods and reorganised harbour organisation. By the mid 1980s, Rotterdam city planners therefore recognised the need for the revitalisation of the run-down dock facilities and the attraction of business to the south side. The ambitious vision projected the regeneration of the Kop van Zuid and the construction of an extensive new business district to meet the demands of the expanding service economy. In 1986, planners called for the adoption of a bold development perspective for the city and Riek Bakker, then Director of the city Department of Urban Development, commissioned Teun Koolhas Associates to provide a design brief for the Kop van Zuid. Consultations with local organisations, including residents' groups, formed the basis of the 1987 planning strategy for the area that included plans for business and housing, the construction of a major new north–south bridge (the Erasmus) over the Maas, a new road, and a metro rail connection and station. By 1996, major achievements included construction of the impressive Erasmus bridge, the metro station at Wilhelminaplein linked to the Wilhelminahof office complex, residential development, construction in the Entrepot area at Zuidkade, and schemes on a variety of other sites.

Partnership and risk

At the time of the research, Piet Rodenburg headed the Kop van Zuid Project Team. Rodenberg (1994) explained the concept of partnership adopted for the Kop van Zuid by referring to partnerships as the most important mechanisms for managing the development of the area. The collaborative approach was backed by central government as a way of establishing partnerships bringing together municipal councils, local and regional businesses, and central government.

Rodenberg argued that the designation of model development projects highlighted a shift in government policy in the mid-1980s. Economic conditions forced the government 'to make drastic cuts in its own budget' even though the 'need for substantial investments in projects for economic, technological, and urban renewal' was a pressing one (Rodenberg, 1994, p. 62). For Rodenberg, this did not imply that central government intended to deflect risk or the financial burdens of development entirely to the private sector. The sharing of risk was important, but partners would work together towards mutually agreed objectives. Partners would retain their independence so that there was a 'contribution of resources and a clear spread of risk to be taken by all parties, as well as a corresponding distribution of the profits' (Rodenberg, 1994, p. 62). There were various interpretations of risk, so specific agreements between parties were necessary to avoid the application of any rigid set of rules. The approach was reminiscent of the British Conservative government's view of the Private Finance Initiative where risk sharing was defined in collaborative projects. The risk factor thus became a critical variable in the determination of relationships between the parties in commercial partnerships.

However, the concept subtly synthesised the Dutch predilection for consensus with partnership. An underlying assumption was that harmonious public–private relationships were essential to achieve success. Collaboration was 'inevitable' where there were mutual interests, and governments and private companies were 'always thrown together' especially in major property deals (Rodenberg, 1994, p. 62). Cooperation depended upon the reconciliation of the commercial interests of companies and governments at the national and local levels. The situation at the Kop van Zuid thus presented the city with a particularly delicate market situation where 'reciprocal rights and obligations in the areas of programme, procedure, and financing' (Rodenberg, 1994, p. 63) were defined by the relationships between the government and the city. There was what Rodenberg referred to as a judicial basis for these relationships embodied in partnership initiatives. This was important because of the Dutch practice of defining the obligations in mutual agreements between parties in urban policy. Contracts between organisations in urban renewal established the terms of agreements and described the roles and obligations of those involved. At the Kop van Zuid, local level contracts ordered relationships and complimented broader agreements between the city and central government. These were found in planning documents that 'convey the municipality's intentions and, in so doing, show how the realisation of the plan is to be implemented' (Rodenberg, 1994, p. 64). In addition a theme in the planning documentation, and evident from research interviews with public officials involved with the Kop van Zuid project, was the desire to stress the importance of the role of the city council. The partnership concept at the Kop van Zuid did not cast the city simply a 'customer, client, or supplier' (Rodenberg, 1994, p. 64). If this had been the case, the city itself would be regarded as an independent 'private party' (Rodenberg, 1994, p. 64) rather than a facilitator for the private sector and a joint partner.

At the Kop van Zuid, four central concepts underpinned the partnership philosophy. These were quality, coherence, synergy and equalisation (Rodenberg, 1994, p. 64). The project was subject to quality control, while its elements formed a coherent mix of commercial and residential activities that would produce profits for the partners. The synergy depended on public–private actions linking neighbourhoods, property developers and the city council in effective and efficient working relationships. The synergy strengthened the political consensus between the parties and established a firm basis for the further development of partnerships. Equalisation spread risks between partners that assumed various responsibilities, although the equal distribution of benefits was measured in social as well as economic terms. For example, communities could benefit even from the most profitable commercial developments if improvements were made to housing and the environment.

Structuring conflict out

Given its complex structure and the variable interrelationships between partners at the Kop van Zuid, risk sharing was difficult to quantify. The notion of equalisation could easily serve as an abstract justification for persuading developers to take part in urban regeneration by paying attention to meeting community needs. Equalisation could be interpreted as risk sharing that served the political purpose of convincing community groups that they stood to gain from projects that benefitted globally competitive corporations or big public sector organisations. Indeed, the social return policy implemented at Kop van Zuid showed how planners and politicians restrained political competition through the close involvement of local community organisations in the formal policy process. The objectives of the Social Return Project therefore reflected a desire by planning officials to address the problem of unemployment in the communities surrounding the Kop van Zuid by recognising the relationships between the Kop van Zuid and surrounding deprived districts south of the River Maas.

A Mutual Benefit Workgroup (1992), established by the city council, defined social return as 'a method of ensuring mutual benefit' for investors in Kop van Zuid and the residents of the nearby districts of Feyenoord, Afrikaanderwijk and Katendrecht. According to the workgroup, it was partly 'thanks to their efforts that the Mayor and Aldermen and the municipal council have taken this step' (Mutual Benefit Workgroup, 1992, p. 4). The Social Return Project stimulated employment, small and medium enterprise development, amenity improvement, and infrastructure enhancement (Mutual Benefit Workgroup, 1992). Small businesses had an especially important part to play by providing employment opportunities in the communities. The corporate private sector could invest in community amenities and social provisions and social, infrastructure, and employment initiatives would raise property and land values thus attracting further investment. The Mutual Benefit Workgroup expected that local residents' organisations would play a major part in supporting the social return strategy. This depended on the backing of the Federation of Residents' Organisations and the endorsement of city councillors. Social return and mutual benefit needed joint planning arrangements between agencies and residents in the making of detailed plans for social housing, job creation, and community improvement. The task of linking the strategic vision of the Kop van Zuid as a focus for international investment together with social return therefore required a 'coalition between parties' with 'diverse interests' (Belderbos, 1994). The involvement of the OBR, district level local government, and city departments meant that they would have to work together effectively to consolidate the social return strategy. Therefore, it was necessary to create clear agreements between the city hall and district council 'so as to avoid deadlock situations occurring' (Mutual Benefit Workgroup, 1992, p. 2).

Group incorporation and political restraint

By 1996, the Kop van Zuid project organisation represented the coming together of community, government and private interests and incorporated neighbourhood groups. The Kop van Zuid project organised rather like a semi-autonomous wing of the city council with contacts with external bodies and a strategic apex at the project level operating through a Project Manager and project Steering Committee. Intricate organisational structures ensured the comprehensive representation of city council departments. The organisation guaranteed that the project as a whole operated along partnership lines and that the management team supported the creation of local partnerships focused around individual urban development schemes and commercial property ventures. The Project Team had responsibility for the implementation of individual initiatives that in 1996 included the construction of the Erasmus Bridge, a metro station, the development of the Entrepot area for recreation and housing, housing at the Landtong site and mixed developments at other locations. The most dramatic evidence of change was at the Wilhelminahof where a huge new complex for the customs office, the tax office and the courts of justice were under construction. Such developments required Project Team involvement with public agencies and developers in different partnerships so designed to implement the individual schemes working with local community groups.

For example, the Bewonersorganisatie Feijenoord (BOF), or Feijenoord Residents' Organisation, worked directly with housing organisations, ethnic minorities, the city council and public officials at the Kop van Zuid. The BOF employed the services of professionals in social housing, planning, and architecture and operated through the Kop van Zuid project, both formally and informally. The BOF was a nonprofit organisation without formal corporate status, but it was 'recognised' by the city council as a legitimate representative of local residents. In 1996, the city council preferred the BOF to have corporate status, but at the time BOF managers wanted to retain maximum flexibility with its prevailing status. In the three neighbourhoods covered by the BOF (Noordereiland, Kop van Zuid and Feijenoord), monthly residents' meetings enabled the organisation to contact local people and disseminate information to the community. The organisation operated at the block level where local interests were directly represented, and the BOF maintained close contact with three other residents' organisations. These were the BOA (in Afrikaanderwijk), the KBO (in Katendrecht), and the BOH (covering Hillesluis) which, together with the BOF, formed a federation of residents' organisations. The relationship with city planning officials on the Kop van Zuid was close enough for the BOF to be consulted about plans at their inception. In 1996, the BOF was a member of three Kop van Zuid working groups including one established to provide advice on the development of the prestigious Entrepot development scheme. The BOF joined such working groups when social housing was an important part of the development strategy in schemes such as Entrepot.

Once involved with a development initiative, the BOF worked as a professionally expert consultancy with the social housing companies and the private sector. The social housing companies often required the advice of the BOF, and there was liaison with local people to make sure that community opinion and professional judgment informed the planning process. Residents enjoyed direct access to the BOF office and often visited there to discuss issues and sort out problems. The office acted as a conduit for the views of residents into the city council and to companies involved in development projects both at the Kop van Zuid and in the surrounding communities. To facilitate this role, the BOF adopted a relatively flat-structured organisation that was not rule-bound and which was easily accessible to local people. According to a BOF representative, over-formal organisation was purposely avoided to improve access and to discourage managerial inertia and bureaucracy. While there was a structured management framework, BOF workers preferred to network extensively, especially at the block level where formality was inappropriate. Neighbourhood groups that constituted the BOF worked within this setting, although a North Island group had developed greater autonomy within the BOF to develop a more distinctively networked approach with maximum managerial flexibility. This experience indicated one of the potential problems with networked organisations occurred when sub-groups sought greater autonomy that could challenge the resource base of the larger network. However, the gains were perceived as adaptability in local organisation, the expansion of community access and sensitivity to local and block level interests.

Outcomes

The Kop van Zuid project therefore combined the development of its prosperous new business district with the provision of social housing and social services for low income groups in surrounding communities. However, the potential for a clash of interests between local residents and planners was ever present despite the consensual approach. This was particularly evident when gleaming new office buildings rose beside ethnic minority communities with high unemployment and social disadvantage. The cooperation of local organisations was therefore crucial in the longer term to contribute to the maintenance of ordered decision making and the effective implementation of city council planning strategies. For public planners, community groups at the local level had to support public policy and have the opportunity to input informed professional opinions to aid the planning process.

Duffy (1998) persuasively argues that the city council's close involvement with the Kop van Zuid project provided a contrast with American style business leadership. The Kop van Zuid did not create business-led partnerships because it relied more upon what Duffy (1998, p. 153) describes as a 'fairly traditional' development model with the private sector regarding partnerships as commercially appropriate. Nevertheless, the project produced many benefits. It fostered close working between the city and the port authority and it provided a stable climate for companies investing in Rotterdam along with the involvement of community interests in detailed development proposals. Also, public planning impacts for specific projects were explicitly elaborated by the city council. Duffy detects strong public accountability in Rotterdam as a result, but the danger was that corporate investors might be less likely to perceive benefits in partnerships that distributed too many benefits publically. Possibly for this reason, and because of negotiating problems, the Kop van Zuid Project Team's (1996) expectations about large scale corporate investment at the 'flagship' Wilhelminapier were initially frustrated. The original development consortium for Wilhelminapier, including SAE of France, MBO of the Netherlands, and Laing of Britain, failed to take up development options (Duffy, 1998) leaving it to MBO to go ahead with projects there. Also, there was disappointing private support for the ambitious business cluster concept as conceived in the early years of the project.

European programme lessons

European Union funding at and around the Kop van Zuid also provided some important lessons for the city council. Rotterdam's experience with an early European Union Urban Pilot project highlighted the organisational learning that had to take place in local programmes before they could be deemed successful. The project came during the transition from a top–down to more of a market sensitive approach to urban policy and it represented one of the European Commission's early ventures in urban regeneration through 'short-life' initiatives, funded for between two and three years. For 1989–93, the ERDF funded thirty-two urban pilot projects throughout the European Union with a commitment of 100 million ECU. The projects focused on economic development, environmental improvements, the revitalisation of historical buildings and sites, and skills and technologies. The objectives of each of the original projects supported by the European Commission, including Rotterdam, were innovative and experimental. In Rotterdam, the freedom provided within the programme provided a contrast to some of the public interventions of the past, and the European Commission encouraged local governments to work in new ways with the private sector in developing imaginative solutions to local level problems.

Rotterdam, in competition with other cities, bid to the European Commission for a programme that contained elements that complimented the Kop van Zuid project and which combined with the Rotterdam Healthy City programme. An objective was to create jobs by establishing enterprises in deprived districts, and the commission approved funding for the programme in September 1990. The bargaining between the City of Rotterdam, the European Commission and central government produced an acceptable agreement that adjusted the original proposals to fit well within European Commission guidelines. A Rotterdam Inner Cities Steering Committee had oversight over the Urban Pilot Project, assuming responsibility for monitoring and evaluation and involving local government, government departments, project managers, agencies and residents. The chair of the committee was from the city council, served by a secretariat managed by the OBR. The OBR claimed that the management structure for the programme 'reflected the local responsibility for the correct execution of this pilot program' (OBR, 1994, p. 43). Officially, therefore, the programme was managed with a structure allowing for the expression of local opinion. However, while this allowed a local view on project design and implementation, the programme was led by public officials working within bureaucratic rules applied by the European Commission. With only limited community group representation on the steering committee, the lack of 'bottom-up' input from local inhabitants became evident. Moreover, a subsequent evaluation report stated that only 'during one meeting the European Commission was directly represented' and that the Dutch Department of Economic Affairs only 'made limited attendance because of the arrangement that the Rotterdam municipality was explicitly responsible for the execution of the project' (OBR, 1994,

p. 43). The report added that 'only in the case of severe problems would central government intervene' (OBR, 1994, p. 43). Because the original objectives for the modestly funded Urban Pilot Project were ambitious, they were not all successfully achieved. This was partly because the programme had to link to a variety of existing policies that committed city council resources to implementing a wider range of complex programmes. The Urban Pilot Project aimed at social renewal that, according to the OBR, was a process aimed at 'pushing back dependence, social exclusion and deprivation, and at creating reasonable living conditions' (OBR, 1994). However, this was a far reaching objective for a programme that was better equipped to deal with very specific local problems where coordination was often difficult to achieve. Nevertheless, the lessons learned provided insights for future strategic thinking and the development of the later European Union URBAN programme in Delfshaven that connected to the Major Cities Policy. It underlined the importance of coordination between programmes and extensive local community involvement in partnerships and multi-agency arrangements.

Policy coordination and group inclusion through the Major Cities Policy

The impetus for reorganising the local delivery of the city's social renewal programmes came with the election of Rotterdam's new city council in 1994. A social renewal project manager, and a small social renewal project team, played an important role in developing decentralised programmes under the supervision of the city council and public agencies. The emphasis was upon the involvement of local people and programmes that reached out to communities. Eleven 'Deltemater' district councils within the city were the basis for the local renewal areas, and the coordination of different city departments contributed to what the city council called a 'social dynamo' with partnerships providing the motive force connecting public and interests to local communities. Programmes combined financial resources from the European Union through the URBAN Programme, the Dutch government, the City of Rotterdam, the district councils and the private sector, although Rotterdam did not have European Objective status.

In July 1995, the Dutch government and the four major cities in the Netherlands signed the Major Cities Policy Covenant. Rotterdam city council recognised the urgent need to deal with crime, drug abuse and other social issues by introducing integrated policies.

Policy integration related initiatives to the development of small and medium sized businesses, measures to reduce unemployment, community safety and the improvement of the 'living environment'. The policy also focused on the local economy and new forms of administrative cooperation. The assumption was that economic growth could boost employment and security and improve the physical environment in the city. To achieve these goals, the policy adopted a 'district-oriented approach' where some initiatives within the programme provided extra resources for existing policy commitments. For example, the policy integrated security, employment, social regeneration and youth policies, and new projects developed under the covenant with central government including the redevelopment of port and business sites. Where possible, initiatives therefore brought together existing activities and projects with the new district-oriented initiatives (City of Rotterdam, 1996). The five districts to benefit from the policy in Rotterdam were each adopted by council aldermen who, with municipal departments, were committed to the effective coordination of central government with the city, private partners and local people. The city council synchronised different policy areas and responsibilities through joint meetings of district councils and local community groups. The aim was for 'social regeneration' using 'the ideas and efforts of residents and other private partners' including shopkeepers' associations, employer groups and housing associations. It was 'only with the contribution and involvement of these groups that the selected areas could intensively be tackled. This contribution was stimulated by means of various working parties, opportunities for comment and discussion, think-tanks, progress controls, and financial contributions' (City of Rotterdam, 1996). At the time of the research, the Major Cities Policy in Rotterdam was successfully established within the areas designated. Project management groups for each area furthered inter-agency cooperation and implemented the strategic vision for the districts. The designated areas thus provided the opportunity for the involvement of the local communities and five municipal aldermen working with the programme who had specialist knowledge of local initiatives. The management groups for each area worked as boards that included the relevant alderman, the chairman of the district council, a city official and an appointed project manager who monitored progress and to supervised the development of local projects.

Adjusting to the market

The contrast between the policy approach in the early 1990s with that at the end of the 1990s was marked. In 1990, the original objectives for the Urban Pilot Project envisaged city council leadership and control through centralised administration. However, the scenarios in chapter eight highlighted the need for a move away from the old approach to the implementation of policies to a blend of public–private action and public administrative reform. The scenarios identified changes in the market that could obstruct policies, but the way around problems was no longer simply by recourse to government. Nevertheless, in practice the reduction of public hierarchy was often difficult to achieve. For example, the economic recession of the early 1990s adversely affected the Kop van Zuid Project and created a cautious climate among private sector investors making it difficult for the city to attract external private sector investment and private involvement in local programmes.

The Major Cities, European and city initiatives drew attention to the economic problems in local communities and the dangers of social tensions and the growth of extremist political organisations. As the scenarios showed, these problems could exist even along with impressive shopping malls and high rise buildings. Creating public–private partnerships under such conditions accented the problems of bringing together groups that sometimes saw the city's future in different ways. The city council and community groups valued participation and consensus through public action, but there remained the problem of achieving stronger private sector involvement in urban regeneration. As Duffy (1998) shows, Rotterdam had not yet developed the high profile business leadership in urban regeneration that existed in the USA. The market demanded radical new policy perspectives that accommodated corporate values, a less regulated investment climate and the need for government to be responsive to the needs of the private sector. Against this, Rotterdam's municipal tradition persisted, as in the Kop van Zuid Project where a perception of city pride originally heavily influenced the strategy. However, the Kop van Zuid put city officials on a steep learning curve that alerted them to the vagaries of the market and the often fickle attitudes of globally active companies. This learning process led the city towards a more realistic view of relationships within public–private partnerships where the stress was upon the mutual interests of partners as opposed to the narrower civic interests of city government. The city council, like the Netherlands as a whole, became better attuned to the market and to the demands of a fast changing economic environment through recognition of the importance of business in economic development. The need for flexible organisational structures and the broader regional context of local economic development policies were also accepted. City officials realised that public–private partnerships required organisational and financial mechanisms suited to market conditions. The dynamics of different partnerships therefore reflected the specific conditions that faced partners and the particular commercial circumstances associated with projects.

The economic development and urban regeneration policy domain therefore was multifaceted. The city council worked within a changing policy process that was adapting to change within several related policy domains. Taken with the requirements of central government in spatial planning, the policy agenda of Municipal Port Management and the needs of business, the city sought policy integration and strategic direction. This was achieved through interdepartmental collaboration within the city and between city and port, and through partnerships. The appointment of a new Minister for Urban Policy and Integration of Ethnic Minorities in 1998 underscored the national coalition government's intention to target issues and develop more integrated policies to deal with the intricacies and complexities of related policy problems. Multiple external contingencies produced policy responses that emphasised joint agency working, decentralisation, efficient allocation of resources and regional competitiveness. This was a policy discourse that would be familiar in Pittsburgh and Birmingham.

10
Evaluation

This book has been about the processes involved in strategy development and partnership organisation. Discussion has avoided the effectiveness and performance of strategies and partnerships. However, this chapter does address the effectiveness issue to the extent that policy evaluation requires the assessment of both policy outcomes and processes. The argument here is that assessing strategic outcome is extremely difficult, but the evaluation of strategy is possible if viewed as part of the wider processes of organisational and political change. The chapter provides an assessment of the varied forms of organisation and politics in the partnerships studied in Pittsburgh, Birmingham and Rotterdam, assuming that such an assessment could form part of an evaluation agenda. This approach therefore does not provide detailed prescriptions for effective management. Instead it suggests a framework that could link an organisational understanding of partnerships to process issues and evaluation.

Effective strategies

Strategies are supposed to improve policy, produce gains for regions and communities, and overcome organisational fragmentation. However, strategies come and go and reassessments often reflect deeper crises in policy domains (Boin and 't Hart, 1998) where there is considerable variety and substantial incoherence in much strategic thinking. Effective strategies should encourage coherence, but as Heidenheimer, Heclo, and Adams (1990) argue, many political values and objectives influence policies, and strategy makers are prone to disagree about purposes and aims. Majone and Wildavsky

(1984) argue that policies therefore have 'multiple meanings' in different contexts and for different political actors.

Despite the problems, the neglect of strategy under risky and unpredictable conditions would be tantamount to admitting that city governments either cannot or should not try to influence events. Formal planning overstates the expectation of control and impedes progress by introducing bureaucratic rigidities into policy even though planners still 'believe organizations should plan' (Mintzberg, 1994, p. 173). For Mintzberg, planners have an almost compulsive desire to fulfill their role as planners. Planning mindsets result from rational management with its emphasis on order and control, and they therefore lead to inflexibility and distortions in policy. When planners defend policies, planning becomes 'a conservative process' that conserves the 'basic orientation of the organization' (Mintzberg, 1994, p. 175). According to this view, planning institutionalises conservatism and reinforces worn-out procedures and structures.

The plea for more intuitive strategy based on skill and learning under changing conditions is appealing because it recognises that policy innovation accompanies change. Intuition with appropriate formal analysis can work, but policy makers should not become preoccupied with even this kind of crafted strategy making. It is better that they have the opportunity to make decisions appropriate to dealing with changing conditions so that they can work in the market to capitalise on the possibilities available to them. Intuition frees policy makers from rigid policy prescriptions and increases their effectiveness as players in the competitive market. However, policy makers in Rotterdam, Pittsburgh and Birmingham do not fully conform to this flexible model because effective intuition is difficult when public officials need solutions that are politically acceptable and where there are formal planning cultures. They therefore use hybrid strategies with differing mixes of responsiveness and formal analysis and different techniques and styles of management. Consequently, Porter's strategic approach is popular in the three case study cities because it implies a continued role for public policy makers. Generic strategies can produce stock responses, but public officials favour strategic rationality despite the problems that this causes in effectively responding to external change. Strategies thus constantly change to realign with reality.

Strategy and outcome

Discussion of the effectiveness of regional strategy is thus problematic at both the macro and micro levels. Macro analysis is about the inherent advantages and disadvantages of different strategic approaches. For example, debate concerns the relevance or otherwise of strategy as a motive for economic development. One view is that strategy is less important for successful economic development than the geographical location of cities. Some cities are simply more fortunate than others by nature of the location in economic growth regions (Harding *et al.*, 1994). The other view is that geographical location has no bearing upon the fortunes of cities. Even those in 'peripheral' locations can prosper if they combine the right growth factors (Chisholm, 1995) and, implicitly, have effective and dynamic political–strategic motivation. The details of this debate are beyond the scope of this book, but consideration of such contextual issues could usefully inform the evaluation of strategic and administrative processes. It could also link to another macro issue concerning the relationship between strategy and market. Imbroscio (1997) refers to strategies and economic models that fail because they do not effectively relate to market conditions. However, even if the market-strategy match is right, policy consistency and coherence are necessary if strategies are to produce desirable outcomes (Rumlet, 1996). But ambiguous policy goals in the case study cities often produced incoherence and vague partnership visions despite community leaders and politicians expectations that policies would respond to local needs and produce measurable benefits and outcomes.

Needs and outcome assessments for whole cities require comprehensive assessments of the factors that affect the condition of urban areas. The European Union Urban Audit is a large scale project that is gathering data on urban issues in 58 cities including Birmingham and Rotterdam. The Urban Audit will provide a macro level comparison of urban change, but using relevant local data for each of the cities. The data relate to socio-economic changes, the nature of civic participation, training and education, the environment, and culture and leisure. These data provide the basis for indicators under each of these headings that will enable the audit team to assess the state of the cities and their progress in developing their economic and social bases (European Commission, 1999).

Nevertheless, grand macro level policy comparisions are difficult given the diversity of national data, differences of territorial definition and the problems of defining baselines for comparison. It may be more fruitful to concentrate evaluation efforts on selected programmes and to identify the factors that make them succeed or fail. At the level of micro analysis, governments and nonprofit organisations assess the outcomes of specific policy interventions, but these may or may not be easily attributable to particular overarching strategies. Governments measure jobs created and new businesses established, but it is difficult effectively to associate these with broad aims when different partnerships and organisations are responsible for different programmes. The Urban Redevelopment Authority in Pittsburgh evaluates programmes by viewing outputs in housing, business development and real estate financing (URA, 1997), and in Birmingham, agency evaluations stress programme objectives and quantified outputs. Local evaluations refer to performance expectations, as in Rotterdam where programmes focus on social return and benefits to ethnic minorities (Anderiesse, Bol, Oudijk and Bons, 1997), but multi agency responsibilities confuse the relationships between aims and outcomes. The Dutch Ministry of Internal Affairs (Hezemans and Weijers, 1996) monitors the performance of the Major Cities Policy nationally, and the Institute for Social and Economic Research at Erasmus University measures local social and economic variables in different cities, but it is local interventions based on effective processes that often matter most in effecting positive changes. Rotterdam's BOF residents' organisation thus encourages local groups to evaluate community initiatives through close involvement with projects. Groups at the Kop van Zuid work as clients with project teams, involving the BOF as an expert participant in the planning process.

In practice it is therefore difficult to specify how particular strategies have affected programmes and their longer term social outcomes unless there is an understanding of a range of change processes. Usually many complex influences on social outcomes make it hard to relate cause and effect, so a possible approach to evaluation is to concentrate more on the processes that influence programmes and lead to particular outcomes (Pawson and Tilley, 1997). In social programmes, evaluators increasingly recognise the need to design programmes in advance more effectively to relate programme aims to desired outcomes by allowing for diversity in partnerships and understanding the relationships between context, strategy, organisation and outcome (Connell and Kubisch, 1997). Shared visions and strategic objectives are important in giving direction to specific programmes, and organisational–institutional characteristics also influence outcomes. Integrating strategy, organisation, and outcome into a theory of how change takes place in regions and local communities could provide a way by which individual partnerships conceptualise their roles within the wider policy setting and partners come to understand their inherent strengths, weaknesses, and capacities (Connell and Kubisch, 1997).

Partnership as process: factors for evaluation

Treating partnerships and programmes in this way highlights a need for the understanding of programmes through evaluations of organisational development and synergy as part of the process of change. In Pittsburgh, Birmingham and Rotterdam, different partnerships therefore provide a variety of organisational contexts for policy implementation because they bring different organisations together and articulate policies and strategic visions in different ways. For that reason it is important to put organisational issues high on the evaluation agenda.

This approach can aid comparison and deepen our understanding of organisational patterning. Partnerships in the three cities superficially have much in common, but a process approach may reveal more. Public officials and politicians commonly recognise the problems of administrative fragmentation and the need to work across traditional local government jurisdictions. However, despite the similarities, comparative study displays important organisational differences at the regional, city and neighbourhood levels. The case study cities have economic development organisations that vary according to the local conditions in each city, and this suggests that contingent factors produce varied patterns of organisation. It also suggests the impossibility of blueprinting strategies to apply under all conditions. Instead, regional economic development and urban regeneration programmes have to take account of local political cultures, institutions, political competition and complex multilevel spatial arrangements.

Other factors also accentuate differentiation. Partnerships in each of the three cities have different developmental histories that produce deeply textured organisational structures and produce varied cultural meanings and processes. For example, in Pittsburgh partnerships developed by building upon the core competencies of established business organisations that assimilated the lessons of many years of cooperation. In Birmingham, political discontinuities between the Conservative and Labour governments disrupted the policy process during the 1980s when business leadership was underdeveloped. In Rotterdam, the Kop van Zuid project represents an uneasy departure from traditional municipal approaches and provides important lessons for policy makers about partnership building under changing conditions. Policy innovations and experimentation not only make market interventions commercially and politically risky, but they also complicate patterns of networking. Governments, companies and nonprofit organisations structure partnerships to increase resources, share risks, reduce political conflict and ease organisational adaptation (Regalado, 1994).

Given such circumstances and the evidence from the cases, at least five important factors are especially relevant when examining the operational and structural dynamics of partnerships. The five factors, detailed below, concern the changing nature of public–private interactions, spatial organisation, organisational development, the achievement of synergy and the effective sharing of risks.

Public–private mix

As part of the process of change, local governments, public agencies, and partnerships develop organisational 'parts' that resemble those in some private sector companies (Mintzberg, 1995b, p. 639). However, Bovaird and Hughes (1995) argue that corporate and local government organisations differ despite often fuzzy boundaries and shared goals. There are parallels between private and public sector restructuring, but also 'differences in the core prescriptions embedded in reengineering and reinventing' (Bovaird and Hughes, 1995, p. 355). For Bovaird and Hughes, government reinvention involves both internal change and the externalisation of relationships through networks and partnerships. In contrast, corporate reengineering stresses internal relationships and process redesign (Davidow and Malone, 1992; Hammer and Stanton, 1995; Bowman and Kogut, 1995). Nevertheless, corporate reengineering still produces reassessments of internal organisational structures and companies join with public and private organisations forming external alliances beyond traditional local government.

This process is complicated because the different spatial levels of public organisations and their different structures all respond in different ways to external environmental changes. City government departments such as education and welfare are also traditionally accountable to elected representatives, but networking challenges the pattern and breaks down old departmental and public–private sectoral boundaries. Local governments have structures that reflect their public service connections, and officials acknowledge the public service ethic that serves as an institutionally stabilising mechanism. Local government officials pride themselves that they serve the public and that they value the views of elected representatives. Public officials operate mainstream services within the constraints imposed by the democratic system, and they cooperate with nonprofit organisations and other local interests. However partnerships mix public and private sector cultures, and local government public officials become more distant from the public gaze when in 'hidden' networks. It is there where public officials and corporate executives identify common interests and visions through political bargaining and compromise.

Spatial coverage

Pittsburgh, Birmingham and Rotterdam illustrate that corporations and city governments adopt confusing definitions of 'the region' particularly when external environments change quickly. Simple demarcations between the regional, city and neighbourhood levels have only a conditional application when explaining networking and organisation, and 'the region' takes on different meanings under different market conditions. There are broad similarities of spatial organisation in the case cities, but partnerships distribute functions to handle subnational issues in different ways according to the circumstances that they confront. The apparent similarities of spatial organisation in the three case study regions are, on closer analysis, less regular than they seem. Superficial similarities between cities demand cautious treatment, as in the Dutch case where public agencies work across and between traditional spatial 'levels,' and where the Rotterdam 'urban region' is more compact than the American. In the Netherlands, American 'regional' spatial categories are sometimes inappropriate given the different scale and interpretation of regional issues. The Dutch frequently adopt a discourse that places the 'region' at the level of the British urban conurbation or the contiguous metropolitan area in the USA. In practice, it is easier directly to compare the local level organisations in each country because at that level functional responsibilities distribute along similar lines. However, even at the local level, spatial likenesses mask wide organisational differences. For example, the Urban Redevelopment Authority in Pittsburgh has citywide responsibilities for economic development in a way similar to the Rotterdam City Development

Corporation (OBR). However, while URA and the OBR operate citywide, their organisations are different. The URA is a public authority and the OBR operates as part of the Rotterdam municipality, and their corporate cultures and developmental histories differ. The URA, the OBR, have some functions and operational roles in common, but they vary in respect to their legal status, organisation, and external connections. The Birmingham Heartlands Development Corporation had a similar policy brief to the URA, but its limited short-life activities targeted only one area of Birmingham.

Organisational development

As indicated in chapter two, contingency theory predicts the variety of organisations that develop in different environments. The analysis emphasises differences in organisations that structure according to the circumstances confronting them. The organisational configuration approach develops this theme by analyzing structural patterns and the external and internal factors that condition organisations and produce dynamic change within them. Partnerships between public and private organisations vary because they are the products of structural adaptation and political bargaining (chapters two and three). Partnerships are products of the adaptation of existing structures and the development of new ones produced by expanding administrative capacities.

City government departments in Pittsburgh, and agencies like the URA, also develop new functional structures and adapt old ones administratively to support local level partnerships. The Pittsburgh 'layer cake' (chapter four) depicts partnerships configuring under changing conditions and transforming themselves as the economy restructures. The Working Together Consortium in Southwestern Pennsylvania developed new structures and programmes to pursue its regional policy aims, but business leaders built upon past collaborative experiences. The Rotterdam Urban Pilot Project adapted existing public administration to support new initiatives, and in Birmingham the city supported partnerships through local government service departments.

Partnerships in the three cities therefore needed either external management support, or they had to develop their own core administrative competencies. The reader will recall from chapter two that Mintzberg's (1996b) six basic parts of modern complex organisations are functional in the development of core competencies. Competencies include the skills of management, the ability to work effectively to attain strategic objectives and factors associated with the delivery of services. In Mintzberg's large complex changing organisation, an operating core delivers the service or product. The strategic apex consists of managers who oversee the running of the organisation, and a middle line of management develops as the organisation grows. With increasing complexity, functional groups develop a technostructure and a support staff. The ideology of the organisation, or its culture, embodies the beliefs and traditions promoted by the organisation. The six parts link through a structure of authority, although the technostructure and support staff generally enjoy greater autonomy than the other parts of the organisation. Extant organisations combine these parts to give different configurations, but organisations also network externally by forming external coalitions that develop into organisational 'sites' for bargaining and negotiation between partners. Clegg (1990) values Mintzberg's analysis of organisations, and Goffee and Scase (1995, p. xvii) also favour the approach precisely because it explains the emergence of 'relatively abstract organizational structures' (Galbraith, 1995). Similarly, Alvesson (1995) provides graphic illustrations that show cultural fields within a framework shaped by interlinking organisational functions.

The case studies reveal structural diversity and competitive power relations in partnerships and the need for partners to establish core competencies that deliver results. Partnerships need to develop functional parts, and these resemble those in single organisations. However, the relationships between the parts of partnerships do not conform to any single structural pattern common to all partnerships. Organisational patterns do not configure watertight types since different organisational forms often coexist side by side and lead to many hybrid forms (Farley and Kobrin, 1995). One configuration in the public sector produces the classic centralised and formalised bureaucracy, like those in bureaucratic departments. In contrast, diversified organisations work in unstable settings that require project teams and management flexibility. The cases show that big city governments have attempted to decentralise services and manage programmes at 'arms length' to achieve greater flexibility and effectiveness. Diversification leads to networking and splits big departments into semi-autonomous divisions that develop their own ways of working and their own organisational cultures. Ironically, policy makers who decentralise frequently pull units together again by adopting hierarchical bureaucratic solutions. It is especially at such times that organisations can be more prone to politics (Rosenthal *et al.*, 1961) as internal conflicts develop during the process of re-shaping the organisation. In multi-member partnerships, political competition can lead to conflict that fragments management authority making difficult strategy, so public policy makers seek holistic, joined-up, comprehensive, shared and commonly owned solutions.

Quinn, Anderson, and Finkelstein (1996) identify different ways in which the managerial 'intellect' and competencies of networks concentrate. The spider's web network (Quinn *et al.*, p. 357) is an informal arrangement with little organisational development where 'the locus of intellect is highly dispersed' (Quinn *et al.*, p. 357) and where there may be little agreement between network members about policy or mission. Possibly no hierarchy or management line exists in a spider's web although these may later develop as the network adopts a stronger mission and role. Quinn *et al.* (1996) also define a 'cluster organisation' that is similar to the spider's web with a nodal structure (see chapter two). For example, many representatives of different sponsor organisations get together to take joint action or discuss policy. Finally, the 'starburst' organisation is like an interorganisational arrangement, 'but for special reasons the organizational units are under some shared ownership' (Quinn *et al.*, 1996, p. 359). Such networks, or partnerships, have highly developed organisational characteristics of their own where the main sponsoring organisations devote resources to administration, operational units and management. This kind of arrangement describes many organisations involved in economic development in Pittsburgh, Birmingham and Rotterdam. It also helps to describe organisations that perform coordinating functions, but with responsibilities 'shared' within a network. The URA in Pittsburgh collaborated with local partnerships, and although it was not itself a partnership, it supported a network of organisations benefitting from URA expertise in urban regeneration. The former Birmingham Heartlands Development Corporation and the Kop van Zuid Project Team had similar roles to the URA in this respect. The Allegheny Conference on Community Development was a classical networked organisation being 'shared' by its business partners with a centralised role as a negotiator of agreements (Scharf, 1997, p. 136). British Single Regeneration Budget Challenge Fund partnerships and American Empowerment Zones and Enterprise Communities also coordinated external partnerships through partnership bodies that linked groups in such ways.

Therefore, there are different degrees of organisational development within partnerships shown in general terms in Figure 10.1. Figure 10.1 shows that partnerships at a particular spatial level can develop strong organisational characteristics while others operate informally. So in Pittsburgh the picture is of sophisticated organisational development with partners sharing risks and organising through webs and starburst-style partnerships with substantial resources going to administration, management and consultancy.

The possible configurations in Figure 10.1 challenge neat categorisations of situations. Figure 10.1 builds upon the analysis provided in chapter two that links politics to organisation, and it rests on the research for the case studies. The axes in Figure 10.1 show different degrees of political competition (see chapter two) and the different degrees of formal organisation linking partners. Figure 10.1 therefore identifies combinations of political competition with forms of organisation. Complex partnership organisation changes over time and competitive relationships produce new configurations through bargaining and compromise. Variability produces a fuzziness and lack of clear distinction between formal networks and decentralised organisation. An emerging partnership therefore can disperse policy making and administrative authority between the partners. The pattern of dispersal depends upon the nature of the arrangements that link the strategic apex with the other parts of the evolving partnership. Figures 10.1 and 10.2 give a qualitative view.

Along the fuzzy boundary between formal organisation and networking lies the 'virtual' organisation (Nohria and Berkley, 1994) with informal networks developing formal patterns of coordination and administration. Local governments often support machine-style hierarchical solutions to maintain financial control, and partnerships that develop substantial administrative workloads and complex interagency cooperation often assume the characteristics of larger public organisations. Figure 10.2 illustrates the degrees to which formal linkages can form as embryonic organisations. The figure shows the tendency for decentralisation in partnerships when there are multiple groups and strong political competition. It also suggests that as politicians and partnership stakeholders attempt to assert control partnerships are more likely to develop hierarchical characteristics. Developed organisation does not necessarily go with centralised hierarchy, but in economic development there appears to be a tendency for the two to reside closely although political tensions can pull apart even the most centralised and highly de-

Figure 10.1 Degrees of Organisation in Partnerships

Degree of Organisation	Organisational Activities	Strategic Direction	Line Management	Professionalism	Culture
Developed Organisation	Complex administrative tasks and formal procedures are developed.	A sophisticated strategic management function exists with management coordination.	Strong coordination and authority structure, possibly with centralisation and hierarchy.	There is strong professional support for managers and policy makers and a defined technostructure. A move towards a more professional approach to dealing with policy problems.	A distinctive corporate culture has developed within the partnership.
Developing and Undeveloped Organisations (producing different degrees of organisation in each case)	Expanding administrative functions may be shared between partners or vested with a specialist partnership body.	Shared strategic vision requires coordination between personnel responsible for oversight of strategy and administration.	There exists developing line management and defined structures and staff loyalties to management.	Professional support develops through consultancy and close relations with sponsors of the partnership. Support for professional approach to policy problems.	There may be a distinctive culture within the partnership, but it still probably reflects aspects of the sponsoring organisations.
Formal Relationships	Administrative tasks tend to be distributed among members of the network or vested with a sponsoring organisation.	Agreement to cooperate may lead to a formal declaration of a shared vision and partnership.	Line management is undeveloped, but potentially partners may seek clearer lines of control.	Some professional support, possibly initially with the drafting of strategic aims.	There is little in the way of a distinctive culture.
Informal Relationships	Few administrative functions or none at all within the network.	No formally agreed strategy. Things are at the ideas stage often with informal meetings.	There is little in the way of a management structure, or none at all within the network.	Growing professional support for joint action from potential partners and other sources can encourage closer working between interests.	The network has no distinctive culture of its own because it is made up of disparate parties.

veloped of organisations. Starburst structures are appropriate for regional partnerships and they can be highly developed organisationally and politically plural. The tighter central control becomes, the less likely a decentred structure will remain intact. Much depends on the political context and the efforts of policy makers within the network to develop line management structures that are suitable for what they want to achieve. The demands and activities of interest groups within the network or partnership can also be influential especially where they challenge any dominant members of the network. Business leadership initiatives tend to starburst because there is an emerging consensus that can support plurality of representation, but if there are too many diverse demands that challenge the consensus, then the network may fragment or become difficult to control. The regional fragmentation in Southwestern Pennsylvania in the 1980s clearly showed the dangers of such over plurality and variations in business leadership. It led to the Working Together Consortium that attempted to organise regional interests by developing key partnership competencies and by providing strategic direction. To that extent, strategy restrained political competition in the region.

Is it possible to develop the contingency model using quantitative measures of political competition and organisation? These variables are difficult to quantify given the lack of coherent and consistent comparative measures of organisational development in economic development. Nonetheless, a very rudimentary approach would be to score degrees of political competition along one axis and score degrees of organisational development along the other. Organisational development could be scored according to values given to the development of the parts of the organisation. This could provide at least a sketchy picture of different configurations, but it would not capture subtle cultural influences, the individual policy visions of groups or complex differences in financial regimes. Additional problems arise when partnerships nest within other partnerships (Perrow, 1986) and when they work at different spatial levels so that configurations are not uniform across levels. For example, the English Government Offices for the Regions network at the national, regional and local levels where economic development and urban regeneration partnerships exist at each level. The quantification of network organisation demands caution to avoid simplistic generalisations that may not apply across spatial levels. If networks operate at multiple levels then, as Perrow (1986, p. 198)

Figure 10.2 Organisational Characteristics of Partnerships at One Spatial Level

	Restrained Competition	Competitive	Conflictual-Dysfunctional
Developed Organisation	This combination produces high organisation and restrained competition. It implies a potential for centralisation and top-down management. Similar to the Kop van Zuid in Rotterdam.	This combination signifies a highly organised partnership with competitive politics. Partners share complex tasks with a strong operating core. Business leadership initiatives in the USA are often like this, and may foster starburst arrangements or be like the early stages of the RDAs in Britain.	Partners are likely to split the partnership. There may be lack of internal control and rivalry between the partners.
Developing Organisation	Partnership has developing organisational structures and restrained competition and hierarchical control (like EU programmes and some SRB initiatives in Birmingham). Can become high organisation.	Partnership can be highly competitive, but there is a strong possibility of a workable compromise if there is an agreement to share risks (like late 1990s regional partnerships in Pittsburgh).	A developing organisational framework for the partnership, but possibly many partners where conflict threatens effective collaboration (like 1980s regional groups in Pittsburgh).
Undeveloped Organisation	Partnership with minimal organisation, but restrained competition. One partner may dominate, or partners could share administrative tasks as the partnership develops.	There are opportunities for effective cooperation if partners strike up a workable compromise and develop effective structures. Spiders webs tend to formalise where partners cohere.	Conflict or lack of agreement threatens to break up the partnership and prevent organisational development and coherence. Similar to Pittsburgh Enterprise Community.
Formal Relationships (preceding a formal partnership)	A network with low, possibly flat organisation, but where there is a general agreement to define the aims of a future partnership.	Diverse actors jostle for position in the context of a debate over proposed partnership aims and objectives. Spiders web network patterns common.	Political disagreements lead to the early break up of the network or informal partnership.
Informal Relationships (preceding a formal partnership)	An informal network where actors retain their own identities but agree on general aims for the future.	Transient, unstable, and unstructured relationships between actors. However, there are attempts to achieve coherence and joint action.	Political conflicts lead to the termination of contacts between actors.

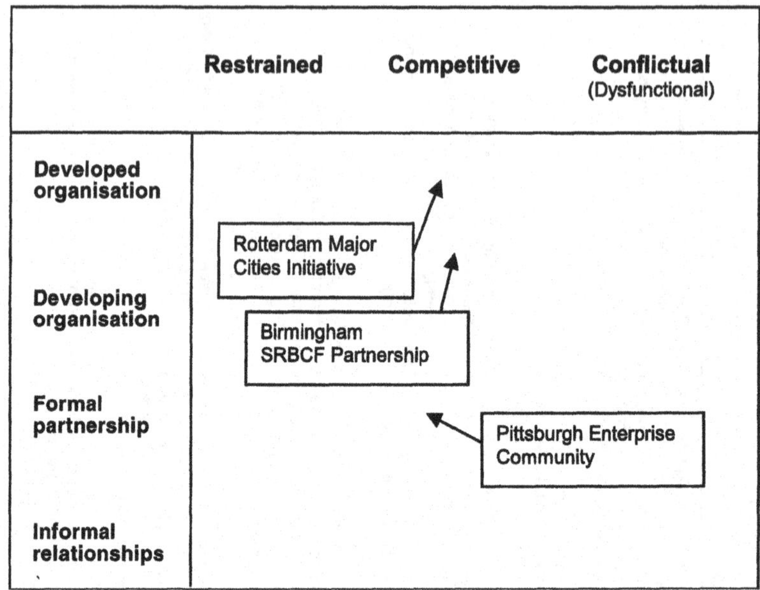

Figure 10.3 Indicative Positions of Selected Local Level Partnerships in the Three Case Study Cities Showing Possible Directional Changes in 1997

suggests, there will be 'different degrees of density in different parts of the network.' Hence, the qualitative approach adopted here.

Pittsburgh, Birmingham, and Rotterdam show that partnerships assume stronger organisational characteristics as administrative units develop, and they have 'trajectories' that change as policy makers alter policies and adopt new strategies. It is possible to draw configuration maps (Figure 10.3) for informal networks and partnerships that track such political and organisational slipperiness. In quite crude terms, Figure 10.3 maps some local level partnerships by plotting their degrees of political competition and organisational development. The figure comes with a health warning that such an assessment of partnerships is tentative and heavily constrained by the limitations of graphical representation. Qualitative assessments and case descriptions of partnerships, taking account of the subtleties of networks and multilevel organisations, could strengthen the analysis. Richly textured partnerships and networks are living, changing, political phenomena that frequently defy the restrictive parameters of two-dimensional representation.

Synergy

Flexibility and innovation in organisation are important because unstable and risky conditions create a need for fast and effective solutions that establish policy control and certainty. This is especially so when social dislocation caused by economic disruption influences community organisations (Quarantelli, 1993). Changing conditions force policy makers to revise commitments and learn from experiences across related policy areas to reduce risk and programme failures. Scott (1995) and Rose (1993) argue that public officials adapt policies and learn from others by creating policy hybrids. Officials often make mistakes, possibly borrowing inappropriate 'pieces' of organisation or 'institutional artifacts' from others (Nisiguchi and Anderson, 1995). However, public policy makers have choices (Sorensen, 1993), and programmes can prevent social fracturing and reduce tensions between conflicting groups if policies are adaptive and based on good practice. Managers in urban programmes therefore stress organisational co-ordination, capacity building and structural innovation to reduce conflict and tensions (Halachmi and Bouckaert, 1995).

In the case study cities, strategy making generates compromises, although partnerships are not always harmonious and are often unsatisfactory for handling intergroup competition (Jarman and Kouzmin, 1994). Policy strategies therefore constantly require modification as public officials seek new solutions to avoid failure and disruption (Jarman, 1994). Success thus depends on groups working together to achieve a synergy. The Pittsburgh–Allegheny Enterprise Community initially failed because disparate partners could not achieve harmony. Attempts to overcome failure led to a willingness to revise policies and develop a new strategy. Local partnerships are often more pluralist because programmes mobilise many representative interest groups, but when policy makers revert to hierarchical management they seek control rather than a healthy synergy. A synergy is especially problematic when groups that dominate the strategy exclude community opinion (Regalado, 1994). Sometimes community activists gain representation on community boards, but the strains between opposing groups can generate conflict and 'tensions between the interior and the exterior of organizations' within the policy domain (Heuvelhof and Bruijn, 1995, p. 175).

Risk sharing

Policy makers try to maximise strategic freedom and often exploit the weaknesses of their competitors (Ohmae, 1982, p. 38). The balance between strategic freedom and control has to be right to reduce conflict over the sharing of operational, financial, and political risks (Bunyan, 1994). Risk sharing is important in economic development in the three case studies because partnerships promote economic growth by mobilising and coordinating divergent interests. Collaboration is economically advantageous because it cements industrial alliances and improves regional competitiveness (Bailey and Shan, 1995).

After 1992, the British government's Private Finance Initiative defined shared risks and responsibilities in urban regeneration (Department of the Environment, 1994). No formal rules for risk sharing existed because 'each area of government policy was different' (Bunyan, 1994, p. 4). But, the government advocated risk analysis and recognised that some risks previously borne by public agencies could transfer to the private sector. Bunyan (1994) identified risks in public–private partnerships that committed large sums of money. In his classification, *procedural risks* were significant when public policy changed or public consents or licenses were unobtainable for projects. The public sector usually bore this risk. *Design risks* were important where the private sector was responsible, say, for the design of a process or design of an asset. *Construction risks* involved cost overruns and defective construction usually regarded as the responsibilities of the private sector. *Maintenance risks* were significant where the private sector took responsibility for the maintenance of an asset over time. *Operating risks* divided between the public and private sectors, being important when an asset failed to operate effectively and did not meet its targets. *Revenue risk* concerned the performance of an asset and the generation of income (Bunyan, 1994). *Financing risk* occurred when the government expected the private sector to carry risks associated with fluctuating financial conditions under which substantial benefits could accrue to the private sector if things went right. Risk sharing between organisations also challenges the distinctions between policy areas such as urban regeneration, social renewal and community policing, and strengthens coalition building and community involvement (Regalado, 1994).

Partnerships in Pittsburgh, Birmingham and Rotterdam thus handle multiple and linked risks in urban regeneration, local economic development, crime prevention and social control (Quarantelli, 1993). Risk sharing can deliver social and economic benefits arising from collaboration. However, Benington (1994) refers to how the economic, social and political dimensions of policies can undermine local autonomy and lead to groups missing the benefits of economic growth. Partnerships therefore can provide the motive force for successful regional economic development, but they are imperfect mechanisms that often exclude community organisations and blunt their political influence (Walzer and Jacobs, 1998).

Conclusions

The evaluation of regional and local governance will therefore be increasingly about the relationship between organisational factors and the processes that produce beneficial outcomes for communities. The message is that the attainment of beneficial outcomes depends on the effective working of regional systems of governance that in the 1990s in the three case cities involve a combination of governmental hierarchy, networking and partnership. This provides a context in which organisational ambiguity accompanies attempts by policy makers to achieve order in the policy process by reducing risks and unpredictable political conflict. Governments and business work together to address the problems posed by a changing external environment and in so doing they form new partnerships and adopt new ways of working. Policy strategies adapt as the external environment changes and as the changing political culture produces new problems and challenges. The governance of regions in each case is thus problematic given the unpredictability of the environment and the constantly shifting strategic perspectives adopted by regional organisations and public agencies. The lack of coherent democratic accountability in the three countries at the regional level, combined with organisational diversity, highlights the problem of establishing viable regional institutions during an era of economic and social change. It also underlines the highly contingent nature of regional governance that implies that traditional institutional reform agendas are inappropriate when it comes to establishing new regional structures responsive to the demands and aspirations of ordinary citizens.

Bibliography

Allegheny Conference on Community Development (1992) *The Future of Our Region*, Pittsburgh: Author.

Allegheny Conference on Community Development (ACCD) (1993) *Toward a Shared Economic Vision for Pittsburgh and Southwestern Pennsylvania*, Pittsburgh: ACCD and Carnegie Mellon University.

Alvesson, M. (1995 edition) *Cultural Perspectives on Organisations*, Cambridge: Cambridge University Press.

Amin, A. (1994) *Post-Fordism: A Reader*, Oxford: Blackwell.

Amin, A. and Thrift, N. (1995) *Globalization, Institutions, and Regional Development in Europe*, Oxford: Oxford University Press.

Amsterdam City Council (1994) *A City in Progress: Physical Planning in Amsterdam*, Amsterdam: Amsterdam City Council.

Anderiesse, R., Bol, P., Oudijk, C. and Bons, C. (1997) *Case Studies of Research and Policy on Migrants in Cities: Rotterdam*, Utrecht: European Research Center on Migration and Ethnic Relations.

Andersen, S. S. and Eliassen, K. A. (eds) (1993) *Making Policy in Europe: The Europification of National Policy-Making*, London: SAGE.

Andeweg, R. B. and Irwin, G. A. (1993) *Dutch Government and Politics*, London: Macmillan.

Audit Commission (1991) *Urban Regeneration and Economic Development: The European Community Dimension*, London: Audit Commission.

Bachrach, P. and Baratz, M. S. (1962) 'Two Faces of Power,' *American Political Science Review*, 56, 947–52.

Bailey, E. E. and Shan, W. (1995) 'Sustainable Competitive Advantage Through Alliance,' in E. Bowman and B. Kogurt (eds), *Redesigning the Firm*, New York: Oxford University Press.

Baldassare, M. (ed.) (1994) *The Los Angeles Riots: Lessons for the Urban Future*, Boulder: Westview Press.

Barnes, W. R. and Ledebur, L. C. (1998) *The New Regional Economies: The US Common Market and the Global Economy*, Thousand Oaks: SAGE.

Belderbos, F. (1994) 'Social Return: Kop van Zuid, Not an Island in Rotterdam South,' in J. van den Bout and E. Pasveer, *Kop van Zuid*, Rotterdam: Uitgeverij 010 Publishers.

Benington, J. (1994) *Local Democracy and the European Union: The Impact of Europeanization on Local Governance*, Commission for Local Democracy (CLD) Research Report, No. 6., London: CLD.

Berg, L. van den, Braun, E. and Meer, J. van den (1997) *National Urban Policy in The Netherlands*, Rotterdam: European Institute for Comparative Urban Research.

Bianchini, F. and Parkinson, M. (eds) (1993) *Cultural Policy and Urban Regeneration*, Manchester: Manchester University Press.

Birmingham City Council (1994) *Round One SRB Bid Document*, Birmingham: Birmingham City Council.

Birmingham City Council (1997) *Birmingham Economic Strategy: Report of Conference Proceedings*, 18 July 1997, Birmingham: Birmingham City Council.

Birmingham City Council (1998a) *Birmingham: The Vision*, Birmingham: Birmingham City Council.

Birmingham City Council (1998b) *Birmingham: Action Plan*, Birmingham: Birmingham City Council.

Birmingham Economic Information Centre (1996) *Survey of Company Head Offices Based in Birmingham, Special Summary* (unpublished), Birmingham: Birmingham City Council.

Birmingham Heartlands Development Corporation (BHDC) (1997) *Annual Report and Accounts, 1996–97*, Birmingham: BHDC.

Bledsoe, T. (1993) *Careers in City Politics: The Case for Urban Democracy*, Pittsburgh: Pittsburgh University Press.

Boin, A. and Hart, P. 't (1998) *Institutional Crises in Policy Sectors: An Exploration of Characteristics, Conditions and Consequences*, Paper for the Conference of the Netherlands Interuniversity Institute of Government.

Bons, C. P. (1994) *Migrants and Neighbourhood Management in Rotterdam*, Rotterdam: Erasmus University.

Bovaird, T. and Hughes, R. (1995) 'Reengineering Public Sector Organizations: A Case Study of Radical Change in a British Local Authority,' *International Review of Administrative Sciences*, 61, 355–72.

Bowman, E. and Kogut, B. (eds) (1995) *Redesigning the Firm*, New York: Oxford University Press.

Bradshaw, M. (1988) *Regions and Regionalism in the United States*, Houndmills: Macmillan.

Brennan, A., Rhodes, J. and Tyler, P. (1998) *Evaluation of the Single Regeneration Budget Challenge Fund: A Partnership for Regeneration*, London: Department of the Environment, Transport, and the Regions.

British Urban Regeneration Association (BURA) (1997) *The Future of Cities: Creating and Delivering a Sustainable Vision: Conference Report*, London: BURA.

Bunyan, D. (1994) 'The Government's Private Finance Initiative: Text of Presentation to the British Urban Regeneration Association (BURA)', London: BURA.

Castells, M. (1977) *The Urban Question*, Cambridge, Mass.: MIT Press and Edward Arnold.

Castells, M. (1989) *The Informational City: Information Technology, Economic Restructuring, and the Urban and Regional Process*, Oxford: Blackwell.

Castells, M. (1996) *The Information Age: Economy, Society, and Culture. Volume 1, The Rise of the Network Society*, Oxford: Blackwell.

Chisholm, D. (1997) 'No Magic Bullets: Privatization's Threat to Urban Public Administration,' *Journal of Contingencies and Crisis Management*, 5, 140–53.

Chisholm, M. (1995) *Britain on the Edge of Europe*, London: Routledge.

City of Pittsburgh (1996) *Development Solutions for a Changing Market*, Pittsburgh: City of Pittsburgh.

City of Rotterdam (1992) *Rotterdam City Plan: A View on the Spatial Development of Rotterdam between 1995 and 2005*, Rotterdam: City of Rotterdam.

City of Rotterdam (1996) *Major Cities Policy*, Rotterdam: City of Rotterdam.

Clark, T. N. (ed.) (1994), *Urban Innovation: Creative Strategies for Turbulent Times*, Thousand Oaks: SAGE.

Clark, T. N. and Hoffmann-Martinot, V. (eds) (1998) *The New Political Culture*, Boulder: Westview Press.

Clarke, M. and Prior, D. (1998) *City Pride in Birmingham: An Experiment in Urban Policy and Governance*, Birmingham: University of Birmingham Paper.

Clarke, S. E. (1993) 'The New Localism: Local Politics in a Global Era,' in E. G. Goetz and S. E. Clarke (eds), *The New Localism*, Newbury Park: SAGE.

Clegg, S. (1990) *Modern Organisations: Organisation in the Postmodern World*, London: SAGE.

Commonwealth of Pennsylvania (1988) *Urban Redevelopment Law (Amended)*, Harrisburg: Commonwealth of Pennsylvania.

Connell, J. P. and Kubisch, A. C. (1997) *Applying a Theory of Change Approach to the Evaluation of Comprehensive Community Initiatives: Progress, Prospects and Problems*, Washington D.C.: The Aspen Institute.

Coyle, D. J. (1997) 'A Cultural Theory of Organizations,' in J. Ellis and M. Thompson (eds), *Culture Matters: Essays in Honor of Aaron Wildavsky*, Boulder: Westview Press.

Dahl, R. A. (1961), *Who Governs?*, New Haven: Yale University Press.

Dahl, R. A. (1982) *Dilemmas of Pluralist Democracy*, New Haven: Yale University Press.

Dahl, R. A. and Lindblom, C. E. (1976) *Politics, Economics, and Welfare*, Chicago: University of Chicago Press.

Daugbjerg, C. and Marsh, D. (1998) 'Explaining Policy Outcomes: Integrating the Policy Network Approach with Macro-Level and Micro-Level Analysis,' in D. Marsh (ed.), *Comparing Policy Networks*, Buckingham: Open University Press.

D'Aveni, R. A. (1994) *Hypercompetition: Managing the Dynamics of Strategic Maneuvering*, New York: The Free Press.

Davidow, W. H. and Malone, M. S. (1992) *The Virtual Corporation: Structuring and Revitalizing the Corporation of the 21st Century*, New York: Harper Business.

Davis, H. (ed.) (1996) *QUANGOs and Local Government: A Changing World*, London: Frank Cass.

Department of Health (1998) *Partnership in Action: New Opportunities for Joint Working between Health and Social Services*, London: Department of Health.

Department of the Environment (DoE) (1994) *Private Finance Initiative*, London: DoE.

Department of the Environment, Transport, and the Regions (DETR) (1997) *Building Partnerships for Prosperity: Sustainable Growth, Competitiveness, and Employment in the English Regions*, Cm 3814, London: The Stationery Office.

Department of Trade and Industry (1998) *Our Competitive Future: Building the Knowledge Driven Economy*, Cm 4176, London: The Stationery Office.

Dicken, P. (1992) *Global Shift: The Internationalisation of Economic Activity*, London: Paul Chapman Publishing.

Dileleman, F. M. and Musterd, S. (eds) (1992) *The Randstadt: A Research and Policy Laboratory*, Dordrecht: Kluwer.

Dodge, W. R. (1996) *Regional Excellence: Governing Together to Compete Globally and Flourish Locally*, Washington D.C.: National League of Cities.

Dowding, K. (1991) *Rational Choice and Political Power*, Aldershot: Edward Elgar.

Dowding, K. (1995) 'Model or Metaphor? A Critical Review of the Policy Network Approach,' *Political Studies*, 43, 136–58.

Dowding, K., Dunleavy, P., King, D. and Margetts, H. (1993) 'Rational Choice and Community Power Structures: A New Research Agenda,' Paper presented to the Annual Meeting of the American Political Studies Association.

Downs, A. (1967) *Inside Bureaucracy*, Boston: Little, Brown.

Dudley, G. and Richardson, J. (1996) *Promiscuous and Celibate Ministerial Styles: Policy Change, Policy Networks, and UK Roads Policy*, Essex Papers in Policy and Government, No. 107, Colchester: Department of Government, University of Essex.

Duffy, H. (1998) 'Partnerships in Two European Cities,' in N. Walzer and B. D. Jacobs (eds), *Public–Private Partnerships for Local Economic Development*, Westport, Connecticut: Praeger.

Dunford, M. and Kafkalas, G. (eds) (1992) *Cities and Regions in the New Europe: The Global–Local Interplay and Spatial Development Strategies*, London: Belhaven Press.

Dunleavy, P. (1991) *Democracy, Bureaucracy, and Public Choice*, London: Harvester Wheatsheaf.

Dunleavy, P. and O'Leary, B. (1987) *Theories of the State: The Politics of Liberal Democracy*, Houndmills: Macmillan.

Economic Development Council of Northeastern Pennsylvania (EDCNP) (1992) *Regional Economic Development: A Strategic Action Plan for 1992 and Beyond*, Pittston, PA: EDCNP.

Economic Development Council of Northeastern Pennsylvania (EDCNP) (1993) *Profile of the Economic Development Council of Northeastern Pennsylvania, May 1993*, Pittston, PA: EDCNP.

Eldersveld, S. J., Stromberg, L. and Derksen, W. (1995) *Local Elites in Western Democracies: A Comparative Analysis of Urban Political Leaders in the United States, Sweden, and the Netherlands*, Boulder: Westview.

European Commission (1994) *Competitiveness and Cohesion: Trends in the Regions*, Brussels: European Commission.

European Commission (1995a) *Cohesion and the Development Challenge Facing the Lagging Regions*, Brussels: European Commission.

European Commission (1995b) *The Implementation of the Reform of the Structural Funds in 1993*, Brussels: European Commission.

European Commission (1998) *Sustainable Urban Development: A Framework for Action*, Brussels: European Commission.

European Commission (1999) *The European Union Urban Audit*, http://www.inforegio.org/urban/audit/index.html

Farley, J. and Kobrin, S. (1995) 'Organizing the Global Multinational Firm,' in E. Bowman and B. Kogurt (eds), *Redesigning the Firm*, New York: Oxford University Press.

Ferman, B. (1996) *Challenging the Growth Machine: Neighborhood Politics in Chicago and Pittsburgh*, Lawrence: University Press of Kansas.

Fletcher, C. (1998) 'Working Capital,' *Pittsburgh Prospects*, August–September, 18–24.

Ford Foundation (1996) *Perspectives on Partnerships*, New York: Ford Foundation.

Franzini, C. S., Smith, R. M. and Frakt, S. B. (1994) 'Rebuilding Camden, New Jersey,' *Economic Development Commentary*, 17, 4, 24–9.

Freeman, J. L. (1955) *The Political Process*, New York: Doubleday.

Friedmann, J. (1995) 'Where we Stand: A Decade of World City Research,' in P. L. Knox and P. J. Taylor (eds), *World Cities in a World System*, Cambridge: Cambridge University Press.

Fuchs, E. R. (1992) *Mayors and Money: Fiscal Policy in New York and Chicago*, Chicago: The University of Chicago Press.

Galbraith, J. R. (1995) *Designing Organizations: An Executive Briefing on Strategy, Structure, and Process*, San Francisco: Jossey-Bass Publishers.

Garreau, J. (1991) *Edge City: Life on the New Frontier*, New York: Doubleday.

Genus, A. (1995) *Flexible Strategic Management*, London: Chapman and Hall.

Giddens, A. (1998) *The Third Way: The Renewal of Social Democracy*, Cambridge: Polity Press.

Glazer, N. (1994) 'Divided Cities, Dual Cities: The Case of New York,' in S. Dunn (ed.), *Managing Divided Cities*, Keele: Ryburn Publishing in association with The Fulbright Commission.

Godfroij, A. J. A. (1995) 'Public Policy Networks: Analysis and Management,' in W. J. M. Kickert and F. A. van Vaught (eds), *Public Policy and Administration Sciences in the Netherlands*, London and New York: Prentice Hall and Harvester Wheatsheaf.

Goetz, E. G. (1993) 'The New Localism From a Cross-National Perspective,' in E. G. Goetz and S. E. Clarke (eds), *The New Localism: Comparative Urban Politics in a Global Era*, Newbury Park: SAGE.

Goetz, E. G. and Clarke, S. E. (eds), (1993) *The New Localism: Comparative Urban Politics in a Global Era*, Newbury Park: SAGE.

Goffee, R. and Scase, R. (1995) *Corporate Realities: The Dynamics of Large and Small Organizations*, London: Routledge.

Gore, A. (1997) *Businesslike Government: Lessons Learned from America's Best Companies*, Washington D.C.: US Government Printing Office.

Gouillart, F. J. and Kelly, J. N. (1995) *Transforming the Organization*, New York: McGraw-Hill.

Government Office for the West Midlands (GO–WM) (1994) *The European Structural Funds, The English West Midlands Program, Objective 2, 1994–96*, Birmingham: GO-WM.

Government Office for the West Midlands (GO–WM) (1996) *Working to Win: A Framework for Competitiveness in the West Midlands*, Birmingham: GO–WM.

Gray, C. (1994) *Government Beyond the Centre: Subnational Politics in Britain*, Houndmills: Macmillan.

Greater Pittsburgh Chamber of Commerce (GPCC) and Pittsburgh Regional Alliance (PRA) (1998) *The Dawn of a New Era*, Pittsburgh: GPCC and PRA.

Grossman, Howard J. (1994a) *Economic Development is Workforce Development: Regional Partnerships*, Pittston, PA: The National Conference of Urban Economic Development.

Grossman, Howard J. (1994b) *Team NEPA – 21st Century*, Pittston, PA: Economic Development Council of Northeastern Pennsylvania (EDCNP).

Grub and Ellis Company (1995) *Corporations Review 1995: Pittsburgh Region*, Pittsburgh: Grub and Ellis Company.

Gurr, T. R. and King, D. S. (1987) *The State and the City*, Chicago and London: University of Chicago Press and Macmillan.

Haffner, R. C. G. and Berden, K. G. (1998) *Reforming Public Enterprises – Case Studies: The Netherlands*, Paris: Organization for Economic Cooperation and Development.

Halachmi, A. (1995) 'Reengineering and Public Management: Some Issues and Considerations,' *International Review of Administrative Sciences*, 61, 329–41.

Halachmi, A. and Bouckaert, G. (1995) 'Reengineering in the Public Sector,' *International Review of Administrative Sciences*, 61, 323–27.

Hall, P. (1995) 'Towards a General Urban Theory,' in J. Brotchie, M. Batty, E. Blackely, P. Hall and P. Newton (eds), *Cities in Competition: Productive and Sustainable Cities for the 21st Century*, Melbourne: Longman.

Hamel, G., and Prahalad, C. K. (1994) *Competing for the Future: Breakthrough Strategies for Seizing Control of Your Industry and Creating the Markets for Tomorrow*, Cambridge, Mass.: Harvard Business School Press.

Hammer, M. and Champy, J. (1993) *Reengineering the Corporation: A Manifesto for Business*, New York: Harper Business.

Hammer, M. and Stanton, S. A. (1995) *The Reengineering Revolution: The Handbook*, London: Harper Collins.

Harding, A. (1995) 'Elite Theory and Growth Machines,' in D. Judge, G. Stoker and H. Wolman (eds), *Theories of Urban Politics*, London: SAGE.

Harding, A., Dawson, J., Evans, R. and Parkinson, M. (1994) *European Cities Towards 2000: Profiles, Policies, and Prospects*, Manchester: Manchester University Press.

Harvey, D. (1988 edition) *Social Justice and the City*, Baltimore: Johns Hopkins University Press.

Hay, C. (1998) 'The Tangled Webs We Weave: The Discourse, Strategy and Practice of Networking,' in D. Marsh (ed.), *Comparing Policy Networks*, Buckingham: Open University Press.

Heclo, H. (1978) 'Issue Networks and the Executive Establishment,' in A. King (ed.), *The New American Political System*, Washington D.C.: American Enterprise Institute.

Heidenheimer, A. J., Heclo, H. and Adams, C. T. (1990) *Comparative Public Policy: The Politics of Social Choice in America, Europe, and Japan*, New York: St. Martin's Press.

Heijden, K. van der (1996) *Scenarios: The Art of Strategic Conversation*, Chichester: John Wiley.

Hendriks, F. (1996) 'Democratic Institutions and the Mobilization of Bias: Comparing Urban Policies,' in J. J. Hesse and T. A. J. Toonen (eds), *European Yearbook of Comparative Government and Public Administration*, Baden-Baden: Nomos Verlagsgesellschaft.

Hendriks, F. and Toonen, T. A. J. (1995) 'The Rise and Fall of the Rijnmond Authority: An Experiment with Metro Government in the Netherlands,' in L. J. Sharpe (ed), *The Government of World Cities: The Future of the Metro Model*, Chichester: John Wiley and Sons.

Heuvelhof, E. F. ten and Bruijn, J. A. de (1995) 'Governing: Structure and Process-Contingent Interventions,' in W. J. M. Kickert and F. A. van Vaught (eds), *Public Policy and Administration Sciences in the Netherlands*, Prentice Hall and Harvester Wheatsheaf: London and New York.

Hezemans, M. L. and Weijers, Y. M. R. (1996) *Jaarboek 1995 Grote-Stedenbeleid*, Rotterdam: Institute for Socio-Economic Research (Erasmus University) for the Ministry of Internal Affairs.

Hunt, D. M. (1997) *Screening the Los Angeles 'Riots': Race, Seeing and Resistance*, Cambridge: Cambridge University Press.

Ietto-Gillies, G. (1992) *International Production: Trends, Theories, Effects*, Cambridge: Polity Press.
Imbroscio, D. L. (1997) *Reconstructing City Politics: Alternative Economic Development and Urban Regions*, Thousand Oaks: SAGE.
Jacobs, B. D. (1993) 'Riots in Britain and the United States: The Bureaupolitics of Crisis Management and Urban Policy,' *Journal of Contingencies and Crisis Management*, 1, 152–63.
Jacobs, B. D. (1996) 'A Bureaupolitical Model of Local Networks and Public–Private Partnerships: Responses to Crisis and Change,' *Journal of Contingencies and Crisis Management*, 4, 133–48.
Jarman, A. M. G. (1994) 'Context and Contingency in Public Sector Disaster Management: A Paths Model of the US Space Transportation System Failure, 1968–1988,' *Journal of Contingencies and Crisis Management*, 2, 191–204.
Jarman, A. M. G. and Kouzmin, A. (1994) 'Creeping Crises, Environmental Agendas, and Expert Systems: A Research Note,' *International Review of Administrative Sciences*, 60, 399–422.
Jessop, B. (1994) 'Post-Fordism and the State,' in A. Amin (ed.), *Post-Fordism: A Reader*, Oxford: Blackwell.
Jessop, B. (1997) 'The Entrepreneurial City: Re-Imaging Localities, Redesigning Economic Governance, or Restructuring Capital?,' in N. Jewson and S. MacGregor (eds), *Transforming Cities: Contested Governance and New Spatial Divisions*, London: Routledge.
Jones, M. (1995) 'TEC Accountability on the Agenda,' *Working Brief*, 64, 1–4.
Jones, M. (1997) 'Spatial Selectivity of the State? The Regulationist Enigma and Local Struggles over Economic Governance,' *Environment and Planning A*, 29, 831–64.
Jones, M. R. and Ward, K. G. (1997) 'Crisis and Disorder in British Local Economic Governance: Business Link and the Single Regeneration Budget,' *Journal of Contingencies and Crisis Management*, 5, 154–65.
Judge, D. (1995) 'Pluralism,' in D. Judge, G. Stoker, and H. Wolman (eds), *Theories of Urban Politics*, London: SAGE.
Kantor, P. (1995) *The Dependent City Revisited: The Political Economy of Urban Development and Social Policy*, Boulder: Westview Press.
Katznelson, I. (1981) *City Trenches: Urban Politics and the Patterning of Class in the United States*, New York: Pantheon Books.
Keating, M. (1991) *Comparative Urban Politics: Power and the City in the United States, Canada, Britain, and France*, Cheltenham: Edward Elgar.
Kickert, W. J. M., Klijn, E.-H. and Koppenjan, J. F. M. (1997) 'Introduction: A Management Perspective on Policy Networks,' in W. J. M. Kickert, E.-H. Klijn and J. F. M. Koppenjan (eds), *Managing Complex Networks: Strategies for the Public Sector*, London: SAGE.
Kickert, W. J. M. and Koppenjan, J. F. M. (1997), 'Public Management and Network Management: An Overview,' in W. J. M. Kickert, E.-H. Klijn and J. F. M. Koppenjan (eds), *Managing Complex Networks: Strategies for the Public Sector*, London: SAGE.
Kickert, W. J. M. and Vaught, F. A. van (1995) *Public Policy and Administration Sciences in the Netherlands*, London: Prentice Hall–Harvester Wheatsheaf.
Kincaid, J. (1994) 'Governing the American States,' in G. Peele, C. Bailey,

B. Cain and G. Peters (eds), *Developments in American Politics 2*, London: Macmillan.

King, D. S. (1987) 'The State, Capital, and Urban Change in Britain,' in M. P. Smith and J. R. Feagin (eds), *The Capitalist City*, London: Basil Blackwell.

King, D. S. and Pierre, J. (eds) (1990) *Challenges to Local Government*, London: SAGE.

Klijn, E.-H. (1997) 'Policy Networks: An Overview,' in W. J. M. Kickert, E.-H. Klijn and J. F. M. Koppenjan (eds), *Managing Complex Networks: Strategies for the Public Sector*, London: SAGE.

Knoke, D., Pappi, F. U., Broadbent, J. and Tsujinaka, Y. (1996) *Comparing Policy Networks: Labor Politics in the US, Germany, and Japan*, Cambridge: Cambridge University Press.

Kop van Zuid Project Team (1996) *Kop van Zuid: City of Tomorrow*, Rotterdam: Kop van Zuid Project Team.

Kouzmin, A. and Jarman, A. (1989) 'Crisis Decision Making: Towards a Contingent Decision Path Perspective,' in U. Rosenthal, M. T. Charles and P. 't Hart (eds), *Coping with Crises: The Management of Disasters, Riots, and Terrorism*, Springfield Illinois: Charles C. Thomas.

Kresl, P. K. and Gappert, G. (eds) (1995) *North American Cities and the Global Economy: Challenges and Opportunities*, Thousand Oaks: Sage.

Kreukels, A. (1992) 'The Restructuring and Growth of the Randstad Cities: Current Policy Issues,' in F. M. Dieleman and S. Musterd (eds), *The Randstad: A Research and Policy Laboratory*, Dordrecht: Kluwer.

Leach, S. Davis, H. and Associates (1996) *Enabling or Disabling Local Government*, Buckingham: Open University Press.

Le Gales, P. and Lequesne, C. (eds) (1998) *Regions in Europe*, London: Routledge.

Leonardi, R. (1995) *Convergence, Cohesion and Integration in the European Union*, New York: St. Martin's Press.

Lindblom, C. E. (1983) 'Comment on Manley,' *American Political Science Review*, 77, 384–6.

Lijphart, A. (1975) *The Politics of Accommodation: Pluralism and Democracy in the Netherlands*, Berkeley: University of California Press.

Logan, J. and Molotch, H. (1987) *Urban Fortunes: The Political Economy of Place*, Berkeley: University of California Press.

Loos, J. van der (1993) 'Physical Regeneration and Employment,' in M. Sudarskis and M. Edwards (eds), *Urban Regeneration in European Cities: Its Physical, Social and Economic Dimensions*, The Hague: International Urban Development Association.

Lubove, R. (1995) *Twentieth Century Pittsburgh: Government, Business, and Environmental Change, Volume 1*, Pittsburgh: University of Pittsburgh Press.

Lubove, R. (1996) *Twentieth Century Pittsburgh: The Post-Steel Era, Volume 2*, Pittsburgh: University of Pittsburgh Press.

Mackie, T. and Marsh, D. (1995) 'The Comparative Method,' in D. Marsh and G. Stoker (eds), *Theory and Methods in Political Science*, Houndmills: Macmillan.

Majone, G. and Wildavsky, A. (1984) 'Implementation as Evaluation,' in J. L. Pressman and A. Wildavsky, *Implementation: How Great Expectations in Washington are Dashed in Oakland*, 3rd edn, Berkeley: University of California Press.

Manley, J. F. (1983) 'Neopluralism: A Class Analysis of Pluralism I and Pluralism II,' *The American Political Science Review*, 77, 368–83.

March, J. G. and Olsen, J. P. (1989) *Rediscovering Institutions: The Organizational Basis of Politics*, New York: The Free Press.

Marsh, D. (ed.) (1998) *Comparing Policy Networks*, Buckingham: Open University Press.

Marsh, D. and Rhodes, R. A. W. (eds) (1992) *Policy Networks in British Government*, Oxford: Clarendon Press.

Massey, D. (1994) *Space, Place, and Gender*, Cambridge: Polity Press.

Mawson, J. and Spencer, K. (1997) 'The Government Offices for the English Regions: Towards Regional Governance?,' *Policy and Politics*, 25, 71–84.

Mayer, M. (1992) 'The Shifting Local Political System in European Cities,' in M. Dunford and G. Kafkalas (eds), *Cities and Regions in the New Europe: The Global–Local Interplay and Spatial Development Strategies*, London: Belhaven Press.

Mayer, M. (1994) 'Post-Fordist City Politics,' in A. Amin (ed.), *Post-Fordism: A Reader*, Oxford: Blackwell.

Mayes, D. (1995) 'Conflict and Cohesion in the Single European Market: A Reflection,' in A. Amin and J. Tomaney (eds), *Behind the Myth of European Union: Prospects for Cohesion*, London: Routledge.

Meyer, J. W. and Scott, W. R. (1992) *Organizational Environments: Ritual and Rationality*, Newbury Park: SAGE.

Middlemas, K. with Crowe, C., Peucker, H., Algieri, F., Badiello, L., Ballester, R. and Griffiths, R. T. (1995) *Orchestrating Europe: The Informal Politics of the European Union 1973–95*, London: Fontana Press.

Miller, D., and Friesen, P. H. (1984) *Organizations: A Quantum View*, Englewood Cliffs, N.J.: Prentice-Hall.

Mintzberg, H. (1994) *The Rise and Fall of Strategic Planning*, New York: The Free Press.

Mintzberg, H. (1995a) 'Five Ps for Strategy,' in H. Mintzberg, B. J. Quinn and S. Ghoshal (eds), *The Strategy Process: European Edition*, London: Prentice Hall.

Mintzberg, H. (1995b) 'The Diversified Organization,' in H. Mintzberg, B. J. Quinn, and S. Ghoshal, *The Strategy Process: European Edition*, London: Prentice Hall.

Mintzberg, H. (1996a) 'Crafting Strategy,' in H. Mintzberg and B. J. Quinn, *The Strategy Process: Concepts, Contexts, Cases*, New Jersey: Prentice-Hall.

Mintzberg, H. (1996b) 'The Structuring of Organizations,' in H. Mintzberg and B. J. Quinn, *The Strategy Process: Concepts, Contexts, Cases*, New Jersey: Prentice-Hall.

Mintzberg, H. and Quinn, B. J. (eds) (1996) *The Strategy Process: Concepts, Contexts, Cases*, New Jersey: Prentice-Hall.

Monkkonen, E. H. (1995), *The Local State: Public Money and American Cities*, Stanford: Stanford University Press.

Morgan, G. (1986) *Images of Organization*, Thousand Oaks: SAGE.

Morgan, D. R. and England, R. E. (1996) *Managing Urban America*, New Jersey: Chatham House.

Mutual Benefit Workgroup (1992) *Social Return: Plan of Action 1992, Translating Investments at Kop van Zuid into Work*, Rotterdam: Mutual Benefit Workgroup.

Netherlands Bureau for Economic Policy Analysis (1997) *Challenging Neighbours: Rethinking German and Dutch Economic Institutions*, Berlin: Springer.

Netherlands Scientific Council for Government Policy, (NSCGP), (1990) *Reports to the Government 37, Institutions and Cities: The Dutch Experience*, The Hague: NSCGP.

Nisiguchi, T. and Anderson E. (1995) 'Supplier and Buyer Networks,' in E. Bowman and B. Kogut (eds), *Redesigning the Firm*, New York: Oxford University Press.

Niskanen, W. A. (1971) *Bureaucracy and Representative Government*, Chicago: Aldine-Atherton.

Nohria, N. and Berkley, J. D. (1994) 'The Virtual Organization: Bureaucracy, Technology, and the Implosion of Control,' in C. Heckscher and A. Donnellon (eds), *The Post-Bureaucratic Organization: New Perspectives on Organizational Change*, Thousand Oaks: SAGE.

Ohmae, K. (1982) *The Mind of the Strategist: The Art of Japanese Business*, New York: McGraw-Hill.

Ohmae, K. (1990) *The Borderless World*, New York: Harper Collins.

Ohmae, K. (1995) *The End of the Nation State: The Rise of Regional Economies*, London: Harper Collins.

Orfield, M. (1997) *Metropolitics: A Regional Agenda for Community and Stability*, Washington D.C. and Cambridge Mass.: The Brookings Institution Press and the Lincoln Institute of Land Policy.

Organization for Economic Cooperation and Development (OECD) (1998) *Managing Across Levels of Government, Country Report: The Netherlands*, Paris: OECD.

Osborne, D. and Gaebler, T. (1992) *Reinventing Government*, Reading, MA: Addison-Wesley.

Ostrom, V. and Ostrom, E. (1997) 'Cultures: Frameworks, Theories, and Models,' in R. J. Ellis and M. Thompson (eds), *Culture Matters: Essays in Honor of Aron Wildavsky*, Boulder: Westview Press.

Pawson, R. and Tilley, N. (1997) *Realistic Evaluation*, London: SAGE.

Peck, J. and Tickell, A. (1994) 'Searching for a New Institutional Fix: The After-Fordist Crisis and the Global–Local Disorder,' in A. Amin, *Post-Fordism: A Reader*, Oxford: Blackwell.

Peele, G., Bailey, C., Cain, B. and Peters, G. (eds) (1994) *Developments in American Politics 2*, London: Macmillan.

Perrow, C. (1996 edition) *Complex Organizations: A Critical Essay*, New York: McGraw-Hill.

Peters, B. G. (1988) *Comparing Public Bureaucracies: Problems of Theory and Method*, Tuscaloosa: The University of Alabama Press.

Peters, B. G. (1997) 'Shoudn't Row, Can't Steer: What's a Government to Do?,' *Public Policy and Administration*, 12, 2, 51–61.

Peters, G. (1998) 'Policy Networks: Myth, Metaphor and Reality,' in D. Marsh, (ed.), *Comparing Policy Networks*, Buckingham: Open University Press.

Peterson, P. (1981) *City Limits*, Chicago: University of Chicago Press.

Pierre, J. (1998) 'Local Industrial Partnerships: Exploring the Logics of Public–Private Partnerships,' in J. Pierre (ed.), *Partnerships in Urban Governance: European and American Experiences*, London and Pittsburgh: Macmillan and University of Pittsburgh Press.

Piore, M. J., and Sable, C. F. (1984) *The Second Industrial Divide: Possibilities for Prosperity*, New York: Basic Books.

Pistor, R., Riechelmann, M., Rijnaarts, P., Slot, L. and Smit, J. (eds) (1994) *A City in Progress: Physical Planning in Amsterdam*, Amsterdam: Dienst Ruimtelijke Ordening Amsterdam.

Pittsburgh Partnership for Neighborhood Development (PPND) (1996) *Pittsburgh Manufacturing and Community Development Networks*, Pittsburgh: PPND.

Pittsburgh University Center for Social and Urban Research (1994) *Economic Benchmarks: Indices for the City of Pittsburgh and Allegheny County*, Pittsburgh: Pittsburgh University Center for Social and Urban Research.

Porter, M. E. (1980) *Competitive Strategy: Techniques for Analyzing Industries and Competitors*, New York: The Free Press.

Porter, M. E. (1985) *Competitive Advantage: Creating and Sustaining Superior Performance*, New York: The Free Press.

Quarantelli, E. L. (1993) 'Community Crises: An Exploratory Comparison of the Characteristics and Consequences of Disasters and Riots,' *Journal of Contingencies and Crisis Management*, 1, 67–78.

Quinn, B. (1996) 'Strategies for Change,' in H. Mintzberg and B. J. Quinn (eds), *The Strategy Process: Concepts, Contexts, Cases*, New Jersey: Prentice Hall.

Quinn, B. J., Anderson, P. and Finkelstein, S. (1996) 'New Forms of Organizing,' in H. Mintzberg and B. J. Quinn (eds.) (1996) *The Strategy Process: Concepts, Contexts, Cases*, New Jersey: Prentice-Hall.

Randstad Cooperation Economic Affairs (ROEZ) (1996) *The Randstad: Key Figures, 1996–97*, Amsterdam: ROEZ.

Regalado, J. A. (1994) 'Community Coalition Building,' in M. Baldassare (ed.), *The Los Angeles Riots: Lessons for the Urban Future*, Boulder: Westview Press.

Regional Economic Revitalization Initiative (RERI) (1994) *The Greater Pittsburgh Region: Working Together to Compete Globally*, Pittsburgh: RERI.

Rhodes, R. A. W. (1992) *Beyond Westminster and Whitehall: The Subcentral Governments of Britain*, London: Routledge.

Rhodes, R. A. W. (1997) *Understanding Governance: Policy Networks, Governance, Reflexivity and Accountability*, Buckingham: Open University Press.

Rijnmond Inter Municipal Agency (1994) *Kaderwet Bestuur in Verandering Hoofdlijnen*, The Hague: Ministry of Internal Affairs.

Rodenberg, P. (1994) 'Kop van Zuid as Key Project,' in J. van den Bout and E. Pasveer (eds), *Kop van Zuid*, Rotterdam: Uitgeverij 010 Publishers.

Rose, R. (1991) 'Comparing Forms of Comparative Analysis,' *Political Studies*, 39, 446–62.

Rose, R. (1993) *Lesson-Drawing in Public Policy: A Guide to Learning Across Time and Space*, New Jersey: Chatham House.

Rosenthal, U., Hart, P. 't, and Kouzmin, A. (1991) 'The Bureaupolitics of Crisis Management,' *Public Administration*, 69, 211–33.

Rosenthal, U. and Roborgh, R. (1995) *Administrative Reform in the Netherlands: Continuity and Change*, Leiden: University of Leiden Paper.

Rotterdam City Development Corporation (OBR) (1994) *The Rotterdam Inner Cities Program: The European Community and Economic and Social Renewal in Rotterdam*, Rotterdam: OBR.

Rotterdam City Development Corporation (OBR) (1996) *Four scenarios: An Economic Vision for the Rotterdam Region*, Rotterdam: OBR.

Rotterdam Department of Urban Planning and Housing (1992) *Rotterdam City Plan: A View on the Spatial Development of Rotterdam Between 1995 and 2005*, Rotterdam: Mayor and Aldermen of the City of Rotterdam.

Rumlet, R. R. (1996) 'Evaluating Business Strategy,' in H. Mintzberg and J. B. Quinn, (eds.), *The Strategy Process: Concepts, Contexts, Cases*, New Jersey: Prentice-Hall.

Sabatier, P. A. (1993) 'Policy Change Over a Decade or More,' in P. A. Sabatier and H. C. Jenkins-Smith (eds), *Policy Change and Learning: An Advocacy Coalition Approach*, Boulder: Westview Press.

Sabatier, P. A. and Jenkins-Smith, H. C. (1993), 'The Advocacy Coalition Framework: Assessment, Revisions, and Implications for Scholars and Practitioners,' in P. A. Sabatier and H. C. Jenkins-Smith (eds), *Policy Change and Learning: An Advocacy Coalition Approach*, Boulder: Westview Press.

Sassen, S. (1995) 'On Concentration and Centrality in the Global City,' in P. L. Knox and P. J. Taylor (eds), *World Cities in a World System*, Cambridge: Cambridge University Press.

Savitch, H. V. and Thomas, J. C. (eds) (1991) *Big city politics in transition*, New York: SAGE.

Savitch, H. V. and Vogel, R. K. (eds) (1996) *Regional Politics: America in a Post-City Age*, Thousand Oaks: SAGE.

Sbragia, A. M. (1996) *Debt Wish: Entrepreneurial Cities, US Federalism, and Economic Development*, Pittsburgh: University of Pittsburgh Press.

Scharpf, F. W. (1997) *Games Real Actors Play: Actor-Centered Institutionalism in Policy Research*, Boulder: Westview Press.

Schneider, M., Teske, P. with Mintrom, M. (1995), *Public Entrepreneurs: Agents for Change in American Government*, Princeton: Princeton University Press.

Schockman, H. E. (1996) 'Is Los Angeles Governable? Revisiting the City Charter,' in M. J. Dear, H. E. Schockman and G. Hise (eds), *Rethinking Los Angeles*, Thousand Oaks: Sage.

Scott, A. J. (1998) *Regions and the World Economy: The Coming Shape of Global Production, Competition, and Political Order*, Oxford: Oxford University Press.

Scott, W. R. (1992) 'Introduction: From Technology to Environment,' in J. W. Meyer and W. R. Scott (eds), *Organizational Environments: Ritual and Rationality*, Newbury Park: SAGE.

Scott, W. R. (1995) *Institutions and Organizations*, Thousand Oaks: SAGE.

Skelcher, C. (1998) *The Appointed State: Quasi-Governmental Organizations and Democracy*, Buckingham: Open University Press.

Smidt, M. der (1992) 'A World City Paradox: Firms and the Urban Fabric,' in F. M. Dieleman and S. Musterd (eds), *The Randstad: A Research and Policy Laboratory*, Dordrecht: Kluwer.

Smith, M. P. (1979) *The City and Social Theory*, New York: St. Martin's Press.

Social Exclusion Unit (1998) *Bringing Britain Together: A National Strategy for Neighbourhood Renewal*, London: The Stationery Office.

Soja, E. W. (1996) *Thirdspace: Journeys to Los Angeles and Other Real-and-Imagined Places*, Oxford: Blackwell.

Sorensen, R. J. (1993) 'The Efficiency of Public Service Provision: Assessing Six Reform Strategies,' in K. A. Eliassen and J. Kooiman (eds), *Managing*

Public Organizations: Lessons from Contemporary European Experience, Thousand Oaks: SAGE.

Stewart, J. and Stoker, G. (eds) (1995) *Local Government in the 1990s*, Houndmills: Macmillan.

Stewman, S. and Tarr, J. (1982) 'Four Decades of Public–Private Partnerships in Pittsburgh,' in R. S. Fossler and R. A. Berger (eds), *Public–Private Partnerships in American Cities: Seven Case Studies*, Lexington, Mass.: D.C. Heath.

Stoker, G. (1991) *The Politics of Local Government*, London: Macmillan.

Stoker, G. (1995) 'Regime Theory and Urban Politics,' in D. Judge, G. Stoker and H. Wolman (eds), *Theories of Urban Politics*, London: SAGE.

Stone, C. (1993) 'Urban Regimes and the Capacity to Govern: A Political Economy Approach,' *Journal of Urban Affairs*, 15, 1, 1–28.

Stone, C. (1996) 'Urban Political Machines: Taking Stock,' *Political Science and Politics*, XXIX, 3, 446–50.

Stone, C. and Sanders, H. (eds) (1997) *The Politics of Urban Development*, Lawrence: University Press of Kansas.

Sudarskis, M. and Edwards, M. (eds) (1993) *Urban Regeneration in European Cities: Its Physical, Social, and Economic Dimensions*, The Hague: International Urban Development Association.

Swanstrom, T. (1988), 'Semisovereign Cities: The Politics of Urban Development,' *Polity*, 21, 83–110.

Tarrow, S. (1991) 'Aiming at a Moving Target: Social Science and the Recent Rebellions in Eastern Europe,' *PS*, 24, 12–19.

Taub, R. P. (1994 edition) *Community Capitalism: The South Shore Bank's Strategy for Neighbourhood Revitalisation*, Boston, Mass.: Harvard Business School Press.

Thomas, J. C. and Savitch, H. V. (1991) 'Introduction: Big City Politics, Then and Now,' in H. V. Savitch and J. C. Thomas (eds), *Big City Politics in Transition*, Newbury Park: SAGE.

Thompson, J. D. (1967) *Organizations in Action: Social Science Base of Administrative Theory*, New York: McGraw Hill.

Thompson, M. and Ellis, R. (eds) (1997) 'Introduction' in R. Ellis and M. Thompson (eds), *Culture Matters: Essays in Honor of Aaron Wildavsky*, Boulder: Westview Press.

Thompson, M., Ellis, R. and Wildavsky, A. (1990) *Cultural Theory*, Boulder: Westview Press.

Tiebout, C. M. (1956) 'A Pure Theory of Local Expenditures,' *Journal of Political Economy*, 64, 416–24.

Toonen, T. A. J. (1998) 'Provinces Verses Urban Centres: Current Developments, Background, and Evaluation of Regionalization in the Netherlands,' in P. Le Gales and C. Lequesne (eds), *Regions in Europe*, London: Routledge.

Torbijn, P. (1993) 'Urban Renewal and in Particular the Incentives of Dutch Policy,' in M. Sudarskis and M. Edwards (eds), *Urban Regeneration in European Cities: Its Physical, Social and Economic Dimensions*, The Hague: International Urban Development Association.

University of Pittsburgh Center for Social and Urban Research (1995) *Black and White Economic Conditions in the City of Pittsburgh*, Pittsburgh: Center for Social and Urban Research.

Urban Redevelopment Authority of Pittsburgh (URA) (1996) *Development Projects and Financing Status*, Pittsburgh: URA.

Urban Redevelopment Authority of Pittsburgh (URA) (1997) *Annual Report, 1996*, Pittsburgh: URA.

US Department of Housing and Urban Development (HUD) (1996) *America's New Economy and the Challenge of the Cities: A HUD Report on Metropolitan Economic Strategy*, Washington D.C.: HUD.

US Department of Housing and Urban Development (HUD) (1997) *The State of the Cities*, Washington D.C.: HUD.

Vernon, R. (1966) 'International Investment and International Trade in the Product Cycle,' *Quarterly Journal of Economics*, 80, 190–207.

Wadley, D. (1986) *Restructuring the Regions: Analysis, Policy Model, and Prognosis*, Paris: Organization for Economic Cooperation and Development.

Walzer, N. and Jacobs, B. D. (eds) (1998) *Public–Private Partnerships for Local Economic Development*, Westport, Connecticut: Praeger.

Waste, R. J. (1986) *Community Power: Directions for Future Research*, Beverly Hills: SAGE.

West Midlands Development Agency (WMDA) (1995) *Annual Report 1994–5*, Birmingham: WMDA.

Working Together Consortium (1996) *The Greater Pittsburgh Region: Working Together to Compete Globally*, Pittsburgh: Allegheny Conference on Community Development.

World Bank (1993) *The East Asian Miracle: Economic Growth and Public Policy*, Oxford: Oxford University Press for the World Bank.

Wusten, H. van der, and Faludi, A. (1992) 'The Randstad: Playground of Physical Planners' in F. M. Dieleman and S. Musterd (eds), *The Randstad: A Research and Policy Laboratory*, Dordrecht: Kluwer.

Yoshino, M. Y. and Rangan, U. S. (1995) *Strategic Alliances: An Entrepreneurial Approach to Globalization*, Boston, Mass.: Harvard Business School Press.

Young, S., Hood, N. and Peters, E. (1994) 'Multinational Enterprises and Regional Economic Development,' *Regional Studies*, 28, 657–77.

Author Index

Subject Index

Printed in the United States
By Bookmasters